Learn beekeeping for beginners - From beekeeping to honey

How to easily learn the basics of beekeeping, keep bees and produce your own honey in no time at all

Sabine Grass

CONTENT

Introduction

Hello nature lover!

What appeals to you about beekeeping? Is it the prospect of producing your own honey and having the delicious natural product fresh on your breakfast table every morning? Or is it the fascination for the organism "bee", which stands out as a masterpiece of evolution? Maybe it's a mixture of both.

When you hold this book in your hands, you have an exciting journey ahead of you! Perhaps the first bees are already buzzing in your garden and flying from the hive to collect plentiful honey. Or maybe you are just toying with the idea of possibly acquiring a colony.

This book provides you with a solid basic knowledge about beekeeping and the tasks involved, about honey extraction and processing, and ultimately, of course, about the honey bee.

What is the bee, what social structures prevail in a bee colony and how can you as a human being intervene in these structures in order to get an abundant honey harvest? All these and many more questions you will surely be able to answer at the end of this book.

Bees for beginners

When you hold this book in your hands, you have already mastered the first step to your successful beekeeping career. In this first chapter you will learn everything you need to know about beekeeping before you get your own colony. First, we will discuss the bee and the characteristics of this species. Next, the skills and characteristics that a beekeeper should have and learn are explained, and finally, the equipment that you should acquire is presented. So this chapter provides you with all kinds of theoretical knowledge and background information about bees, beekeeping, bee health and also the importance of the bee for a healthy environment.

THE BEE

To date, over 500 species of bees are known to be native to Germany alone. The honey bee is only one of many, but the one that gives us the honey that is so delicious. This chapter is solely about the honey bee. You'll learn what sets the honey bee apart from the other bee species and makes it special - and why exactly it's so important to beekeeping.

The bee from a zoological point of view
Zoology is the study of the animal kingdom and within this study, all animal life forms that walk the earth are

classified in a system. For example, there is the distinction in different classes, such as insects, mammals, fish and reptiles. Here, the bee belongs to the class of insects. Within this class, the honeybee is classified in the order of Hymenoptera. Within this order, the bees form several families, including the family of true bees (Apidae), to which the honey bee belongs with the Latin technical term *Apis mellifera*. In total, there are about 12,000 different bee families worldwide.

What now distinguishes the honey bee *Apis mellifera* most from the other bee families is the common hibernation as a colony. This exists, at least in the bee families native to Germany, only in the honey bees. They join together to form a colony that stores up a winter supply of honey over the summer, which they use to bring themselves and, above all, their queen through the winter healthy and well-fed.

BIOLOGY OF THE HONEY BEE

In order for you to properly understand the bee and respond to its needs, it is useful to first learn about the realities of this animal and get to know it a little better. In this subchapter, the biology of the honey bee will be discussed. You will learn which anatomical and physiological peculiarities distinguish the bee and how the individual bees of a hive communicate with each other.

Relationships

The biggest difference between the honey bee *Apis mellifera* and the other bee species lies in their way of life and the common hibernation in a hive. From a purely external point of view, the bee species differ only slightly from each other. They are therefore all descended from an original "primordial bee".

Bees belong to the phylum of arthropods (Arthropoda). As such, they have legs that are divided into different segments. Other arthropods include scorpions, spiders and crustaceans. In addition, the body of the honeybee shows two distinct constrictions. This results in a three-part division of the adult insect body (also known as the imago) into head, thorax, and abdomen. These constrictions mark the bee as belonging to the class of notch animals (insects). Within this class, bees belong to the Hymenoptera. This includes bees, ants, wasps and other insects with translucent wings. These translucent wings, of which all Hymenoptera have two pairs, are the distinguishing feature of this order.

The order Hymenoptera can be further divided into the plant wasps and the waist wasps. The last group also includes bees. Within the group of the waist wasps the honey bee can be assigned to the stinging wasps and forms here the own superfamily Apoidea. Within the Apoidea several genera develop - among others also the genus of the honey bee *Apis mellifera.*

Mimicry of hover flies

Bees are considered useful and dangerous at the same time because they secrete venom with their sting, which is lethal to very few mammals, but is usually perceived as at least very painful. Some other insects take advantage of this healthy respect by adapting their appearance to that of the more dangerous insects. This phenomenon is called mimicry and the best example is the hoverfly. This has a yellow-brown or yellow-black striped abdomen, mimicking the bee. Do not be fooled by this. The hoverfly is harmless - and useless for honey production.

Wasp waist

The bees belong to the waist wasps and are characterized by a distinct constriction between the chest and abdomen, the waist. This also gave rise to the term wasp waist, which is used today to describe mostly female body shapes.

The waist wasps have another peculiarity: they all have a sting. Here, a distinction can be made between two species in terms of how this stinger is used. The laying wasps (Terebrantia) use their sting to lay eggs, while the stinging wasps (Aculeata), to which the honey bee also belongs, have converted their sting into a defensive sting with a venom gland. This is also the reason that only females have a defence sting - as it has evolved from the ovipositor tube originally used for laying eggs. Other representatives of the stinging wasps are the true wasps, the golden wasps and the digger wasps and also the ants.

Digger wasps and spider wasps still use the sting to stun prey and then transport them to their own nests as food for the larvae.

However, the honey bee and other representatives of the stinging insects only use the sting to defend themselves - which is why it is also called a defensive sting. The venom injected by the sting is designed to cause pain in the attacker, so the plan to harm the bees is quickly abandoned.

Apiformes - the bees

Arrived at our today's honey bees! But not only the bees belong to the Apiformes, but also, for example, the bumblebees. There are solitary, i.e. individually living bees, and also state-forming species. The honey bee is the only German bee species that hibernates as a colony. Most species, however, are so-called solitary bees, which live completely alone and also provide their brood nest independently. There are also, similar to the realm of birds, cuckoo bees. As the name suggests, these are bee species that lay their eggs in other people's brood nests so that they are also supplied there. Some cuckoo bees even eat the eggs and larvae of the host bee, so that it exclusively supplies the cuckoo bee larvae.

Apiformes basically feed only on plant food, which would be: pollen, nectar and honeydew. Pollen, also known as pollen, is produced by the plant for sexual reproduction. Bees use pollen as a protein-rich food source. Pollen differs in surface structure depending on the plant species and thus it is possible to analyze the pollen in honey and accurately assign it to different plant families, genera and even individual species. The teaching of this honey analysis is called *melis-sopalynology.*

Nectar is produced by plants to attract insects, which then take care of spreading the pollen. The bees use this nectar to produce the classic flower honey. Nectar is basically a sugary liquid secreted by special glands called nectaries. Depending on climatic conditions, the prevailing weather and the time of day, the amount of nectar per flower varies, sometimes considerably.

Lastly, bees also feed on honeydew. Honeydew is secreted by small insects that suck on the plants. These are mainly aphids, bark lice and cicadas. Honeydew consists mainly of sugar and is therefore sometimes called leaf honey. The honeydew gives rise to honeydew honey, which is also called forest honey. The nectar and the honeydew are stored in the so-called honey bladder. This is a crop-like structure on the proboscis. The two substances are collected exclusively by the females and used to feed the larvae. Pollen is transported in different ways and accordingly bees can be divided into different species. There are the basket collectors, which include the bumblebees and honeybees. These collect pollen into a pocket on their hind legs. Leg collectors do not have such a pocket, but attach the pollen to the bristle ridges of the hind legs.

Representatives of this species are, for example, fur bees, long-horned bees, saw-horned bees and thigh and trouser bees. The next species is called the abdominal collector. They have a pocket on the abdomen comparable to the cup-gatherers, in which they can safely store the pollen. These include leafcutter bees and mortar bees, for example. Last but not least, there are the crop bees, which do not transport pollen externally, but stow it internally in a crop. Goiters are exclusively solitary bees, such as the native species silk bees and masked bees.

The honey bees

Among honeybees, there is not only the species *Apis mellifera*, the Western honeybee. Instead, 9 honey bee species are known worldwide so far. The Western honey bee is the only native honey bee species in Europe and also the species with the greatest importance for beekeeping. Originally, the Western honey bee comes from Europe, Africa and the Near East, but it has been spread all over the world by humans. Only in Asia is the Eastern honey bee (*Apis cerana*), which is native to that region, also used for beekeeping.

The Western honey bee can be divided into several subspecies, which have been influenced by man through breeding. Thus, about 25 different breeds were created, of which mainly **four breeds** are used **for beekeeping in** Germany:

• **Carnica**: Carinthian bee; *Apis mellifera carnica*

• **Ligustica**: Italian bee; *Apis mellifera ligustica*

• **Mellifera**: Dark European bee; *Apis mellifera mellifera*

- **Buckfast**: breeding hybrid from different breeds

Tip: If you want to acquire your first bee colony, you should choose a colony that comes from the same region where you want to keep it. By mating different bee colonies in close proximity, an own "breed" develops over time, the so-called landrace. This is optimally adapted to the climate and local conditions of its home and usually forms more stable and healthier colonies.

ANATOMY - THE BODY STRUCTURE OF THE HONEY BEE

The weight of a single honey bee is just 0.1 grams. Its anatomy, its body structure, is highly filigree and light so that it can stay in the air without much effort. At the same time, it is stable and robust enough to carry a payload of 50% of its own body weight. A bee can carry up to 0.05 g of nectar and fly several kilometers back to the hive with it.

A true feat of which the bee is capable due to its special anatomy. As an insect, the bee has the classic constrictions and the tripartite division of the body into head (Latin caput), chest (Latin thorax) and abdomen (Latin abdomen).

External anatomy: body shell, wings and legs

A bee faces various problems in the course of its life: It must fly while being light enough to carry transport weight in addition to its own weight. It must be agile and nimble to move around in a confined hive. And it must be able to climb rough terrain and vertical cliffs, in this case mostly honeycombs.

The legs of the bee are attached to the center of gravity together with the wings for better aerodynamics. The body shell is formed by an exoskeleton. This outer shell stabilizes the insect body and gives it its shape. Similar to what the skeleton of mammals does, only from the outside. Muscles and organs all lie within this exoskeleton. In order to be particularly light and thus capable of flight, the structure is made of a composite of chitin fibers with structural proteins. The chitin shell gets its stability from its structure: it is laid out as ribs and folds and is thus resilient and hard, but at the same time particularly light. Chitin is a polysaccharide, a polysaccharide that occurs naturally in fungi, arthropods such as insects and mollusks, and some species of fish. It is comparable to cellulose, which gives plants a certain stability. Chitin, for example, provides a similar stability. This is even significantly higher than the stability of cellulose due to a nitrogen-containing acetamide group.

Legs have the honey bee six pieces in number, all of which are equipped with small claws at the end. The multitude of legs and the barbs allow the bees to safely and effectively cross even the most impassable terrain.

Bees

Different honey bees are called bee creatures, which are anatomically different from each other. In a hive of the Western honey bee, there are three of them: the drones with their particularly large eyes, the queen, which stands out with a significantly longer abdomen, and the workers.

Worker bees, as the prototype of bees, are equipped with universal equipment. They constitute the very largest part of the bee population of a colony and are all female, but with an inactive sexual apparatus. This is the biggest difference between a worker bee and the queen.

The queen looks very similar to a worker from its external anatomy and differs only by the active ovaries and special scent glands. In the mated state, when the production of eggs starts, the abdomen of the queen swells significantly and she can now be easily distinguished from the workers from the outside. The queen's scent glands secrete special pheromones that are important for the colony. They strengthen cohesion and provide the coordination basis for all work in the hive. Pheromones are special scents and attractants and are used for general and special communication. Thus, certain pheromones of the queen can influence the behavior of the bees. The queen uses her pheromones as a tool for conducting her colony.

The drone is the male bee. Its only task is to mate with the queen and thus contribute to the survival of the colony. He has no sting and has larger wings than the workers to fly faster and keep up with the queen. For mating during the nuptial flight, it has special hair pads on its hind legs that enable it to hold onto the queen in flight. In order to perceive the queen's pheromones particularly sensitively, the drones' antennae are longer and equipped with more sensory cells than those of the workers. And probably the most striking feature is the drone's large eyes, which are accompanied by excellent vision, so that the drone can perceive the queen bee without doubt even at a great distance and go about its work.

The caput of the honey bee - head
The first part of the tripartite body is the head. This is the superior control center and seat of most of the sensory organs. Insects do not have a brain as known from mammals and humans. Instead, several large nerve nodes, so-called ganglia, are located in the head, which together have the function of a brain. Outgoing and incoming nerve tracts ensure that the bee always knows exactly what is going on in its body and can consciously control all muscle movements.

Two large compound eyes are located on the side of the head. These each consist of several thousand individual eyes (called ommatidia), which enables the bee to see complex shapes and movements. Each ommatidium consists of a lens and sensory cells, so it is a tiny eye that is fully functional on its own. The eyes are immobile, perceive only a small part of the environment and send the information to the ganglion center, where the information from all ommatidia is assembled. Thus, many individual images become a grid that represents the environment. With compound eyes, bees can perceive movement very well and even in fast flight, the environment is displayed sharply, but with some loss of detail. Details can only be displayed by the compound eyes if they are in the immediate vicinity.

Frontally on the forehead are three single eyes, which are called point eyes (ocelli). The ocelli can only be seen when looking very closely at the bee, their diameter is only a few millimeters. They are not even the size of a pinhead and are also half hidden by the bristles on the bee's head. The ocelli each contain several hundred light-sensing cells. By the large lenses also smallest light quantities are bundled, which underlines the purpose of the Ocellen as light perception organ. The bee has a particularly reliable internal clock due to its ocelli.

The antennae also originate from the side of the head. Laymen like to call them antennae, but the term antennae is anatomically correct. The antennae enable the bees to smell and feel and to perceive vibrations all around. The antennae are built up in short segments, 10 links in number. Each segment consists of a short tube, which is very thinly coated with chitin. Inside lie countless nerve cells for sensory perception. In addition, vessels lie inside the segments that contain the hemolymph, the insect's blood, as well as small tracheae that belong to the tracheal system, which is part of the insect's respiratory system.

In addition to the sensory cells and sensory pits distributed on the antennae, there are also small tactile bristles for perceiving environmental stimuli and communicating with each other. The bees can actively move the antennae independently of each other and thus use them universally. They are also important for social exchange. Here, the nurse bees palpate the queen's body, and during social feeding, the workers maintain contact via the antennae. If this contact breaks off, social feeding also ends.

The drones have a special feature in their antennae. They have one more limb than the workers and the queen. The antennae of the drones each consist of eleven limbs. Thus the drones are particularly equipped to perceive the pheromone of the queen. Another distinguishing feature is that the antennae of the male drones do not have tactile bristles.

The bee does not have a mouth, as known from mammals. Instead, it has a proboscis, which serves to absorb nectar and honeydew as well as water. The honeydew is not sucked up directly with it, but dabbed off. For this purpose, the bee uses the tip of the proboscis, which is called a spoon. The proboscis forms another distinguishing feature between queen and worker bees, in the latter it is longer so that it can also reach nectar sitting at the very bottom of the flower. Bees can taste excellently with their proboscis, which contains numerous chemical sensory cells. Furthermore, the proboscis is also involved in social feeding (trophallaxis). Here it picks up the drop of food presented by the worker bee between her mandibles. In addition to the proboscis, she has two jaws, the mandibles. With these, she can pick up solid components of her food - the pollen. The mandibles are also used for wax molding for honeycomb construction and for cleaning the brood cells.

The bee has several glands on its head that act either exocrine or endocrine. Exocrine glands secrete secretions to the outside. The salivary gland, the mandibular glands and the foraging glands are exocrine glands of the bee. Endocrine glands release their secretions inward into the track of the hemolymph. Here the bee has two endocrine glands named *Corpora cardiaca* and *Corpora allata.*

Hemolymph is a fluid that roughly corresponds to the blood that flows through the veins of mammals and birds. Its function is to transport nutrients, hormones and breakdown products. One major difference from blood is that hemolymph carries very little oxygen, which is why it does not require blood pigment. Instead, the distributi- on of oxygen takes place via the tracheal system.

The thorax of the honey bee - breast
This is the second segment in the insect body. In the bee, it is the segment with the largest muscle mass. Here are the flight muscles, which occupy the largest space in the thorax in terms of space. The flight muscles are indirectly working muscles - they are not directly connected to the wings. Between them is the chitinous carapace, the exos- keleton. The muscles attach to said exoskeleton from the inside and deform it so that the wings move accordingly on the outside. The wings are hinged to the exoskeleton, which leads to a leverage effect that amplifies the muscle

movements even more, since small movements on the exoskeleton of the thorax translate into large wing movements.

Bees possess two pairs of flight muscles. The dorsoventral muscles run from dorsal (the dorsal side of the bee's thorax) to ventral (the ventral side of the bee's thorax) and, when contracted, cause the thorax to contract slightly. This is translated into a flapping of the wings.

In addition, there is the pair of longitudinal muscles that run longitudinally in the thorax - from front to back. If these muscles contract, the thorax narrows in its transverse diameter and bulges out somewhat. This is translated into a flapping of the wings.

Due to the articulated connection of the wings to the thorax, it is possible for the bee to attach them to the body. In the attached state, a contraction of the dorsoventral muscles and the longitudinal muscles is not translated into a wing beat. However, since muscle movement always produces heat, bees can produce heat by actively moving their flight muscles in the hive without actually flying. This type of heat regulation in the hive is specifically used by the bees in the winter months and in spring when brood begins.

As a hymenopteran, the bee has two pairs of wings, the forewings and the hindwings. In flight, however, it can link them together by a kind of Velcro, so that both wings act aerodynamically as a single wing. The bee can voluntarily break this link, for example, when the wings are unused on the body and also for more precise maneuvering when entering the hive or approaching a flower. The wings are constructed in a particularly light way. So-called webs of chitin ensure that the wings remain stable, and the delicate, translucent chitin skin stretches between these webs. In the cocoon, the wings are folded small. Immediately after hatching, they unfold and the chitin webs fill with air. Shortly thereafter, the webs stiffen, giving the wings their stability.

The pattern of webs and nodes on the wing surface allows differentiation of different bee breeds. The webs divide the wings into individual cells, which can be divided into radial cells, cubital cells and discodial cells. Honey bees have three cubital cells, from their length ratio one can form the cubital index. It is this index that differs per bee breed.

The legs of the honeybee are laid out in three pairs - as are the pairs of legs of all adult insects. Since each pair of legs covers specific tasks, they are usually distinguished by small anatomical features. In addition to locomotion, the legs also serve bees for climbing. To gain a foothold on slippery surfaces, bees have small adhesive lobes on each claw limb. Basically, the legs are constructed in different segments: Coxa (hip), Trochanter (thigh ring), Femur (thigh), Tibia (splint), Tarsus (foot). The tarsus, in turn, consists of several limbs. The claw limb with the adhesive lobe is called the pretarsus in the bee.

The front legs, the first pair of legs, are also used for preening and especially the builder bees use it very regularly when building the combs. They use it to pass small balls of wax from the abdomen further forward to the mandibles. For preening, there are preening embrasures on the heel and a special spike on the tibia (splint). This arrangement creates a small opening that is used to remove foreign bodies from the antennae. Especially often these are small pollen grains that stick the sensory cells of the antenna.

The hind legs, the third pair of legs, are particularly suitable for differentiating between workers, drones and queen bees, as special anatomical features are located here depending on the task. The worker bees have their collection pockets for pollen, also called panties, on this pair of legs. The drone has special bristles as adhesive pads on the hind legs, with which it holds on to the queen in flight so that it can carry out mating undisturbed.

The abdomen of the honey bee - abdomen
From the outside the abdomen looks inconspicuous. It is yellow-brown banded and bears, at least in female bees, the weir spine. Inside, it provides space for most of the internal organs of the bees. Accordingly, there are small openings on the abdomen of the bee, called stigmata. The respiratory system has openings on the abdomen, as do the anus, the sex organs and special glands.

The exoskeleton of the abdomen is segmental in structure, the individual chitinous plates being muscularly interconnected. This makes the abdomen particularly mobile and elastic. The bee can therefore bend its abdomen in all directions - not least in order to always use its defense stinger exactly where it needs it. Workers and the queen have six abdominal segments, drones have seven segments. The individual segments consist of a dorsal plate (the tergite) and the abdominal plate (the sternite). These are connected by the flank membranes (interseg-mental membranes). Each segment is interconnected by intersegmental membranes. These elastic intersections between the hard plates of chitin are necessary to allow the abdomen to expand in circumference.

The muscles of the abdomen are used to rhythmical-ly tighten and loosen it and to perform a pumping moti-on. This pumping motion is important for breathing, since the tracheal system can only passively transport air, but does not actively suck it in as a mammal's lungs do. The filling and emptying of the honey bladder also occurs via contractions. The honey bladder serves to store the nectar and honeydew. This is where honey ripening al-ready begins. A part of it the collecting worker gives in the hive, a part she uses for her own nutrition and a third part she uses for social feeding (trophallaxis). All bees in the hive have a honey bladder filled to approximately the same extent. Not only the degree of filling of the honey bladder can change, but also the fat body - an organ roughly comparable to the liver - can vary its size. There-fore, the elasticity of the abdomen is particularly im-portant. The fat body stores carbohydrates and can build up fats from sugar when needed. It is also the supplier of the building blocks the bees need to build beeswax. In the queen, the volume of the abdomen doubles as the ovaries swell to accommodate the massive egg production.

On the underside of the abdomen, the side of the sternite plates, are the wax glands. These are not perceptible in inactive state. They become active in the construction bees. Here the location of the glands can be traced very well. The eight glands in number are located in pairs on the outer side of the third to sixth segment. The glands end in the wax mirrors. These are smooth surfaces arranged in pairs. On these surfaces, the secreted wax droplets harden into wax platelets. This is pushed out backwards. This process is called "wax sweating". Now the finished wax plate is advanced to the first pair of legs and kneaded by the mandibles until it has the desired consistency to be used in new combs, cell wall or cell cover.

The weir stinger of the bee is probably the one body part that everyone - bee lover or not - knows in any case. The sting apparatus is located at the rear end of the abdomen and consists of a venom bladder, the sting muscles, two glands and the sting itself. Due to small barbs on the stinger, it gets stuck in the skin of humans or animals after a sting. The entire venom apparatus is torn out, the venom bladder remains on the stinger and can empty completely into the attacker. A ripped out venom apparatus means certain death for the bee. This happens mostly only when mammals have been stung, because they have thick skin in which the barbs get stuck. If the bee hits another insect with its stinger, the barbs do not take hold and the bee lives on afterwards.

Venom production already starts on the third day of life and is at its peak around the 15th day of life. The venom bladder contains 0.1 mg of venom at this time. Now the worker bee becomes a guard. Once the venom blister is emptied, it does not refill. Since the stinger has evolved evolutionarily from the laying tube, only females possess a stinger.

Finally, Nassanoff's gland should be briefly explained here. It is a pheromone gland used by special trace bees. Tracker bees show other bees the way to new sources of honey or a new hive. The pheromone is released and disseminated by the sterling. To do this, the trace bee stands in front of the flight hole, raises the abdomen and pulls back the last two dorsal scales (tergite). This exposes the excretory duct of Nassanoff's gland. Now the tracker bee flaps its wings violently and the specific scent of its own colony mixes with the pheromone from Nassanoff's gland. This results in a colony-specific odor cocktail that the bees can follow.

PHYSIOLOGY OF THE HONEY BEE

Physiology is the study of the processes in the body organs up to the processes in the individual cells. It thus encompasses everything that takes place inside the body. The various regulatory circuits and organ systems work closely together to maintain homeostasis. Homeostasis is the balance that prevails in the body. If this balance gets out of joint, the organism becomes ill and can even die if the body's own regulatory mechanisms do not succeed in correcting the imbalance. Important for this internal homeostasis are, among others, the nervous system, the hormonal system, the digestive tract with the connected internal organs and the pheromones acting on the bee from the outside, which are secreted, for example, by the queen or other bees in the hive.

Food intake and digestion

Like humans, bees need essential nutrients that their organism cannot produce itself. They must therefore obtain these minerals and vitamins from their food. Only vegetarian food comes into question as food for our honey bees. They collect nectar, honeydew and pollen. Also identical to the nutritional physiology of humans, bees need carbohydrates, fats and proteins for their energy metabolism and to build up their own bodies. Pollen is the protein supplier par excellence, while nectar and honeydew contain mainly sugar, i.e. carbohydrates. Depending on the plant species, pollen contains about 20% protein, which consists of individual amino acids, and the large protein macromolecules are broken down by the bees' digestive system into just these individual monomolecules of amino acids. The bee then reuses the released amino acids elsewhere.

The bee needs fats only in insignificant quantities, which are also covered by pollen. Nectar and honeydew do not contain any fats, but this is not a problem for the bee because it can synthesize fats itself in its fat body. To do this, it uses carbohydrates, which can be converted to fatty acids in the fat body via several intermediate steps. These are stored in the body as depot fat and can be converted into energy in times of need.

In addition to carbohydrates, fats and proteins, minerals and vitamins are also vital substances for the bee. They are absorbed through nectar and honeydew, which contain salts and trace elements. Pollen contains vitamins, essential fatty acids and other minerals and trace elements. In addition, pollen contains some secondary plant substances that the plant needs for itself. These secondary plant compounds are often used in naturopathic medicine in humans to achieve various effects, such as anti-inflammation, detoxification and protection against free radicals. The digestion process of the collected nectar and honeydew already begins in the bee's proboscis and crop, where the whole is stored for the return flight. Here, digestive enzymes from the bee's saliva join the mixture of nectar and honeydew and begin breaking down the carbohydrates. In nectar and honeydew, the carbohydrates are mostly present as disaccharides. A disaccharide consists of two simple sugars that are biochemically coupled to each other. The job of the digestive enzymes is to break this biochemical bond. This produces the single sugars, so-called monosaccharides, glucose and fructose. Protein digestion does not begin until the bee's midgut. Here, the pollen is mixed with enzymes designed to break down protein.

Pure vegetarians?

Bees feed almost exclusively on plant products, or rather the excretions of aphids, the honeydew. However, it happens again and again that the larvae in the brood cells get sick and die. These dead larvae are usually - unless a lot of larvae die at once - not simply thrown out of the hive, but serve as a valuable source of protein and are eaten by the workers.

The honey bubble

You have already heard about the crop that bees use as a temporary storage for nectar and honeydew. This crop has the technically correct name "honey bladder." An extension of the esophagus, the honey bladder sits at its end in the front part of the abdomen. This is where nectar and honeydew are stored and mixed with digestive enzymes from the foraging sap glands (the hypopharyngeal glands) so that pre-digestion of the sugar and the ripening process into honey can begin here. The hypopharyngeal glands are located in pairs in the head of the workers, they are hardly developed in the queen and the drones. Their activity reaches the peak during the period as nurse bees. Only during this time the bees produce very high quality secretion for feeding the brood and the queen (the royal jelly).

The honey bladder holds about 0.05 to 0.06 milliliters, which corresponds to about 50% of the honey bee's own weight. They are thus true load carriers. Back in the hive, the bee decides what to do with the contents of its honey bladder. It can digest it (partially) itself and use it as a food source. To do this, it opens a valve at the rear of the honey bladder, the so-called proventriculus, which connects the honey bladder with the midgut. Most often, however, it pumps the contents directly into a honey cell to build up a food supply. The third possibility is to use the juice for social feeding (trophallaxis).

The hormone system

The main component of the hormonal system is the bee's endocrine glands, which secrete hormones within the body and thus ensure their distribution throughout the bee organism. This is done by releasing the hormones into the hemolymph. Hormones are considered messenger substances and mediate various processes within the bee's own body.

The most important hormone glands of the bee are the corpora cardiaca and the corpora allata. The paired corpora cardiaca is located directly behind the brain and can store synthesized hormones and produce its own hormones. The hormones of this gland have the tasks to stimulate other endocrine glands and to regulate molting. The queen has more developed corpora cardiaca than the drone and worker.

The corpora allata are a series of smaller glands located at the side of the esophagus, whose function is to produce hormones for general and larval development until metamorphosis. In the queen, they take over the production of sex hormones, ensure yolk production and maturation of the eggs in the ovaries.

Exocrine glands - venom glands, scent glands, and wax glands.

In addition to the wax and venom glands already described and the pheromone gland (Nassanoff's gland), there are also two exocrine glands, i.e. glands that release their messenger substance to the outside, which are of great importance, especially in the queen.

These are the tergite gland, which produces queen-specific pheromones, and the mandibular glands, which mix together the queen substance. This is a mixture of about 30 different substances. The nurse bees ingest the secretion as they nurse the queen, causing their own ovaries to shrink. This ensures that the queen remains the undisputed head of the bee colony.

Hormones and pheromones - a strictly coordinated system

The hive consists of a large colony of bees, workers of all ages and tasks, drones and a queen. The whole is called the bee. This bee must function excellently to ensure its own survival. So that each individual bee now knows exactly what to do, the hormones and pheromones are indispensably important and provide an excellent and in this complexity so far unique form of communication in the animal kingdom.

In addition to the hormones that act internally and the externally acting pheromones, there are also the so-called kairomones. These are also messenger substances that emanate from the bees but are perceived by other species. A striking example is the varroa mite, a pest of beekeeping. The varroa mites can perceive the kairomones and read from them which brood cells will soon be capped and in which brood cell there is drone brood.

Excrement and urine of the bees

Bees do not have kidneys that filter the blood and fish out metabolic waste products as in mammals. Instead, bees have the malpigic vessels. These are tubes opening in the rectum into which the end products of metabolism are discharged with active transport processes. First and foremost here is urea. Urea is formed during the detoxification of ammonia, which is produced during protein metabolism in every organism. Together with indigestible pollen residues, the undissolved urea is then excreted with the fecal bladder.

The fat body

As you have already read earlier in the anatomy chapter, the fat body is roughly comparable in function to the liver of mammalian organisms. It is a central metabolic organ of all insects and serves as a storehouse for nutrients. Thus, the fat body can take apart substances and use the components to synthesize other substances.

The organ is located in the abdomen, consists of several lobes and is surrounded by hemolymph all around. The lobing increases the metabolically active surface area of the organ, thus increasing efficiency. A large fat body means that there is an abundant supply of food. A small, narrow fat body means that the bee is drawing on its own body reserves. Storage of nutrients occurs mainly in the fall. These are consumed towards spring, with the beginning of the breeding season.

The fat body stores glycogen, a multiple sugar, i.e. a carbohydrate, as well as fats and proteins. The larva also feeds on these stores during metamorphosis, since no external food intake is possible there.

Another important task of the fat body is the production of fats for its own supply. Fat is present in the bee's diet only in traces, so the fat-like precursors needed to build up the beeswax have to be produced by the bee itself. This is done by the fat body producing fatty acids from carbohydrates and then combining these with alcohols to form esters. The builder bees in particular therefore have a very active fat body, as they are busy building around the clock in the period from April to July. During this time, the food supply is usually abundant, so the sugar they need is easy to bring in in the form of nectar and honeydew.

The tracheal system for breathing

Bees do not have lungs, and other insects also completely lack this air-exchanging system of mammals. Also, oxygen is not transported in the blood bound to hemoglobin, but is completely independent of the bees' "blood system" - the hemolymph.

Instead, insects have what is known as the tracheal system for their respiration. These are chitin-reinforced tubes that branch throughout the body and end blindly in the tissues. The four respiratory openings, the stigmata, are located on the right and left sides of the tergites (dorsal plates) of the thorax and abdomen. The stigmata lead through a small cage into a short trachea, which then terminate in an air sac in the abdomen. This means that in the abdomen there is an air sac on the right and on the left, from which the tracheal system branches into the tissues.

Honey bee sensory organs

The senses of the honey bee work partly completely differently than you know it from yourself. Bees perceive many of the impressions more selectively, but all the more clearly, so that a special picture emerges.

Bees have an excellent sense of smell; they are true scent specialists and can perceive even the smallest amounts of pheromones from their own colony and flower scents and store them in their olfactory memory. The bees use these scents to orient themselves and communicate with each other. The necessary sensory cells (the receptor cells) and sensory pits are located in clusters around the mouth, on the antennae and on the feet. Each receptor cell is specialized for the perception of a specific substance.

The movable antennas, which can be oriented in all directions, are perfect for perceiving fleeting scents and locating their origin. For this purpose, they are peppered with olfactory plates, pore plates and membrane plates. Drones have a particularly large number of these chemical olfactory plates in order to clearly perceive the queen's pheromone.

A large number of sensory bristles and sensory pits are also located on the pretarsus, especially of the forefeet, in order to be able to perceive the taste and smell of the flower approached in direct contact with it. Thus, bees can taste the sugar content of a solution through their foot sensory cells. In addition, sensory cells are located in the mouth area, i.e. on the mandibles and the labial palps (mouth palps).

The bees' scent memory is capable of storing odors. The complete scent of the flower is not stored here, but an abstract image of the flower is created by selectively perceiving only individual components of this scent. Thus, information about the sugar content is also included in the memory. This is useful for the next nectar collection, since even flowers of the same plant species can differ greatly in their nectar quality and sugar content. In this way, the bees ensure that they fly to flowers where the composition of the nectar is optimal for them. The compound eyes are specially designed to see movement and produce sharp images within their own movement. Thus, they have good temporal resolution. They are also able to make UV light visible to the bee, and even on cloudy days the bee can determine the position of the sun. The dot eyes, the three little eyes on the forehead, are important for the bees because of their position, presumably for stabilizing flight, and also serve for light-compass orientation and a functioning day-night rhythm.

Due to their ability to detect UV light, flowers look completely different to the bee than they do to humans. Thus, a flower that appears pure white to you may contain a distinct pattern to the bee. This type of vision allows bees to recognize special pollen and sap stains on plants and determine the ideal approach site to quickly collect nectar and pollen.

Bees cannot hear; they have no sensory cells for this. Instead, they detect vibrations in the air and orient themselves by them. The dance language of bees also falls into the vibration category. The tail dance is performed in the hive itself, where it is pitch black. The bees cannot follow the dance with their eyes in the classic way, but use their sense of vibration. Instead, vibrations in the range of 15 Hz (abdomen) and 250 Hz (thorax) are generated and passed on to the other bees by vibrating the honeycombs. These read the vibrations with the subgenual organ and the chordotonal organs. These are sensory organs that are able to pick up the vibrations. The subgenual organ is located at the tibia, the chordotonal organs are located at the joints of the legs, once between femur and tibia and once between tibia and tarsus.

The chordotonal organs are responsible for determining the position of the limb sections in relation to each other and perceiving changes. Thus, the movement of the ground can be read from the position of the joints.

Further sensory cells for the perception of vibration are located on the antennae. The bees use their antennae both to feel, measure and perceive each other. Queens, for example, use them to measure brood cells and decide whether to place a fertilized egg or an unfertilized egg inside. Worker bees use their antennae during social feeding (trophallaxis), among other things.

Last but not least, the bee has cushion-like sensory bristles on its joints and individually arranged sensory bristles on its antenna, which serve to perceive gravity. These provide information about how the individual body parts relate to each other.

COMMUNICATION AND BEHAVIOR

In order to understand your bees from head to toe, it is important that you are familiar with bee behavior and communication among them. Only if you understand and know the processes of the colony and hive well, you can directly detect deviations that indicate disease or other problems and intervene early to help.

There are beekeeping clubs that have set up a show hive. These are hives that are equipped with glass panes, so that you can look inside the hive and observe the bees very well. If you do not have this possibility, you can change the situation a bit in your own hive. Hang only a single comb in the comb hive for observation and then watch your bees at work. You should only do this in ideal weather conditions.

The bee

Often the community of the bee colony is called the bee. This is a very appropriate name, because it usually seems that all bees act as one unit, as one large living organism. So the bee includes all the workers, drones and the queen, as well as the brood. But also all stores and the entire honeycomb construction are counted to the bee, since they indispensably determine the survival of the colony.

Pheromones

As you have already learned, pheromones are scents produced by the bees and released to the outside. They are used to communicate with each other, but they work with a small delay because they have to be distributed throughout the hive and the entire colony after being released. Only when all bees have absorbed the pheromones does the desired behavior become visible as activity in the colony. A short-term exchange of information is not possible via pheromones. The pheromone mixture "queen substance" is brewed by the queen from about 30 individual substances and released through the mandibular glands. In a colony that is too large, the concentration is diluted to such an extent that the worker bees begin to create queen cells. Wise cells are special brood cells that serve to raise a queen. Thus, preparations for swarming or silent re-positioning occur. In addition to the normal pheromones used for daily communication and the colony's own odor, there are certain alarm pheromones that the bees emit when an attack occurs. This notifies the entire colony and puts it on alert. The alarm pheromones are basically released when a bee stings. This also means that the attacker is marked with the alarm pheromone and the bees will respond more aggressively. Once you have been stung, you should not approach your bees that day and change your clothes and wash your skin well to get rid of the alarm odor.

Another type of scent communication is scent marks. Scent marks are left by bees through the tarsal gland, the Arnhardt's gland, where the bees walk over with their feet. So mostly when entering and leaving the hive. The tracker bees continue to use secretion from the stercal glands to mark flowers. The scent from the pheromone and the inherent scent of the flower then produces a characteristic mixture that the other bees can follow in order to find the advertised source of grape.

Trophallaxis - the social feeding
Trophallaxis involves the exchange of food from the honey bladder of one animal to another adult. So the bees feed each other, it has nothing to do with rearing larvae. At the same time, the exchange of food is not the primary goal at all, but the trophallaxis serves mainly for the exchange of information as well as the transfer of the queen substance.

For example, a bee that has just performed the tail dance to communicate where it has found a good food source (tracht) can be literally begged for some food by several bees. In this way, the bees taste the advertised crop and can better locate the crop source based on the taste.

Trophallaxis also occurs more frequently during the winter, in the winter cluster, so that the bees can supply each other with their honey supply, which is largely stored in the honey bladders of all bees. Especially after swarming, until the honeycomb construction has progressed so far that honey can be stored here again, the honey bladders are the only storage place for food. Trophallaxis follows a fixed pattern and is always initiated by a begging bee that contacts another bee with its antennae. If this bee wants to share its food, it folds back its proboscis and retrieves a drop from its honey bladder, which it delivers between its mandibles. The begging bee picks up this drop with its proboscis. As soon as the contact of the antennae breaks off, the food exchange is finished.

Honeycomb

Only through the construction of honeycombs in a new dwelling does a bee swarm become a real bee colony, as they thereby create a brood nest and the prerequisite for honey storage. The construction of new honeycombs is complex and demands a real high performance from the bees. One comb contains about 70 g of wax. In order to produce 1 kg of wax, bees have to expend the same amount of energy and material as they use for 4 to 10 kilos of honey.

New construction is only possible if they also have

this material available, which requires an abundant supply of grapes. Therefore, new construction only takes place in spring and early summer. There must also be a queen in the colony, because only then are the bees stimulated to build.

During the time of new construction, all worker bees help together and produce the wax in their fat body. If the dwelling stands and is not extended, then there is only a small number of construction bees, which take care of small repairs in the honeycomb structure and the covering of the honey cells and the brood cells. For this purpose, however, they often use already existing wax instead of producing new wax.

Sterzeln

Sterzelnde bees help the hive bees to find their way back to the hive by means of a characteristic, hive-specific scent plume. Therefore, this behavior occurs more frequently when young bees set out on their first orientation flights or after the colony has moved into a new dwelling.

During sterling, the bees stand at the flight hole, buzzing their wings and simultaneously lifting their abdomen to expose Nassanoff's gland. The rapid wing beating causes the bee's pheromone scent to mix with the odor from inside the hive, creating an odor that is characteristic and distinctive of the individual colony. This is spread in the environment and shows the returning bees the safe way back to their own hive.

Tail dance - the dance language of bees

Tracker bees have a special task within the bee colony. They are the scouts that set out to identify new sources of honey. Once they have found a good hive, they return to the hive and perform their tail dance to activate the other bees and encourage them to follow them to the new hive source. They use the dance to convey all the information the bees need to find the cluster, distance and direction, yield and type of cluster.

The dance is performed on the honeycomb by the tracker bees in order to make it vibrate. For this purpose, there are special areas of the honeycombs that are not connected to the wall at the outer edge. These vibrate particularly well and facilitate the exchange of messages.

Depending on how well the tracker bee manages to activate the other bees and inspire them with enthusiasm for its new source of honey, the more bees will fly out to it. It is comparable to a competition, in which the bee with the best tracht source wins and directs the most workers there. This ensures that the good sources of honey are flown to preferentially.

Helping bees ensure that all inactive bees in the hive are also aware of the tail dance. These helping bees run through the hive and activate the others by grabbing them with the front and middle pair of legs and generating strong vibration with their abdomen to "wake up" the inactive bee, so to speak, and make it aware of the tail dance taking place.

To represent the direction in which the hive source is to be found, the bees use gravity and the position of the sun as fixed reference points. If the hive source is exactly in the direction of the sun, they do the dance in the vertical from top to bottom. Depending on the deviating angle of the tracht to the sun, the dance angle is adjusted accordingly. And if that is not ingenious enough, the bees also take into account the natural migration of the sun across the sky and calculate the appropriate angle over time, so that they can precisely adjust their tail dance here. The distance from the source of the pollen is represented by the duration of the vibration. Simply put, the rule is that the longer the vibration, the farther away the grapevine source is. The type of the grape source is passed on by trophallaxis. Thus, the bees receive information about the taste and smell of the grape with a sample. The more bees fly out, the more bees activate with their dance further bees to set off to the Trachtquelle. In this way, it is possible to run through an activation cascade within 15 to 30 minutes and to send out almost all of the active collectors to the new, productive source of grape.

The brood

In summer, the life span of a worker bee is 5 to 6 weeks. Therefore, if a colony does not take care of the permanent offspring, it would die out within a few weeks. The queen, cleaner bees and nurse bees are primarily involved in brood care.

Cleaner bees clean the combs from the remains of the previous brood and make sure that the brood cells are prepared so that the queen can lay an egg inside. If the larva hatches from this egg, the nurse bees take care of providing this larva with food sap, later with honey and pollen, until the larva finally develops into an adult bee through metamorphosis.

The nurse bees not only take care of the larvae, but are also responsible for providing the queen with protein-rich food so that she can continue to maintain the high performance of egg laying. For this purpose, she is supplied by the nurse bees with royal jelly, which they mix together in their feeding glands.

Expansion and division of the people
The more food there is, the more the colony brings in and the more the colony can grow. The prerequisite for this is that there is enough space in the hive so that new combs can be built, because more brood cells must be created. The first thing that is needed here is for the builder bees to create brood cells. Here they vary the size, because for the division of the colony it needs enough drones. The drone cells are somewhat larger in diameter than the cells for workers. The queen measures this with her antennae and lays an unfertilized egg in the drone cells and one fertilized egg in each of the cells for the worker bee offspring.

If swarming is imminent or a re-pollination becomes necessary, the builder bees create extra large cells, the so-called queen cells. These always hang vertically from the brood comb and are occupied by the queen with a fertilized egg. As soon as a larva hatches from the egg in the queen cell, it is fed exclusively with royal jelly by the nurse bees, which ensures that a queen develops - a female bee with ovaries.

The swarming - moving out to a new home

Next comes swarming. As mentioned at the beginning, the Western honey bee, which is native to Germany, is the only species of bee that hibernates in the colony. Now, to ensure the survival of the species, it is necessary not only to maintain the bee colonies, but also to reproduce and thus grow. As soon as the new queen in the queen bee cell is about to hatch, the old queen leaves the hive with part of the swarm. They usually do not fly far, but settle nearby in a cluster of swarms. This is a very impressive picture, as the bees seem to literally stick together. Now, trace bees fly out and go in search of a new home. This is the moment when you as a beekeeper should capture the swarm, otherwise it will be lost to you.

Wild bees swarm regularly also for hygienic purposes. If there is no beekeeper to replace the combs from the hive, the combs age with each brood that takes place. Protein-containing cocoon remains and remnants of brood reduce the size of the combs with each passing and provide excellent food and an attractant for wax moths, mites, aphids and other parasites. To avoid lethally high parasite pressure for the colony, swarming and abandonment of the old nest occurs.

The Umweiselung - birth of a new queen

A queen bee has a life expectancy of 4 to 5 years. This is impressively long. But from about three years on, the egg production decreases and the queen begins to age, because slowly the supply of sperm that she has acquired on her nuptial flight is also used up. By this time, the queen has produced about 1.2 to 1.5 million eggs.

At the same time, the production of the pheromone mixture queen substance also stagnates and the information goes around in the colony that a new queen is needed, and the bees put on the queen cell. There are two types of natural re-positioning that distinguish you as a beekeeper. The re-referral with swarms, when the colony divides and two colonies are created, and the silent re-referral, which happens without the beekeeper noticing. Here the queen ages and is replaced by a young, healthy, efficient queen. There is no swarming of a part of the colony, because the bees do not follow the old queen without her strong pheromone signal.

Re-pollination can only take place if drones are available to mate with the new queen. The male bees develop in the period from May to July with the sole task of mating the young queen during the nuptial flight and transferring their seed supply to her. The drone is dependent on the workers, as it does not fly out and thus does not feed itself, but survives only by social feeding or by taking advantage of the colony's honey supply.

Recreation - The loss of the queen

A colony without queen will die after a short time, because the offspring is missing. In nature, the loss of the queen bee sometimes occurs due to various events - this can also happen due to the beekeeper, for example, if an inattentive beekeeper loses the queen during relocation.

To save the colony from certain demise, the nurse bees now become active. A female larva is not fixed during the first three days of its development. If it receives royal jelly, the royal food juice, during the entire period of its development, it becomes the queen. If the nurse bees change the feeding on the third day to the food mash of pollen and honey, a worker bee develops. If the colony now notices the loss of the old queen, they convert brood combs with larvae younger than three days into so-called re-creation cells. Usually, five to six of these are created. The replenishment cells are easily recognizable by their significantly larger dimensions. The prerequisite for a re-creation is the presence of uncovered brood in the hive. Capped brood is already too old and too differentiated and can no longer develop into a queen. Furthermore, drones must be present so that the young queen can go out on a nuptial flight to return fertilized and ready to lay eggs.

THE TRACHT - THE FOOD BASIS OF THE BEE

By "Tracht" is understood the totality of the food supply, which is available to the people and at the same time the current Tracht is subdivided, depending on what is brought in at the moment: Nectar, pollen, honeydew, tree sap and also water are included here. Depending on the time of the year, the harvest varies. For example, at the beginning of the incubation period, more protein is needed, so pollen is preferably collected here. You can tell from your bees by observing them carefully when they enter the hive what kind of tracht they are bringing with them: Pollen and tree sap in the pockets of the hind legs; if the flight is heavy, the honey bladder is full of nectar or honeydew.

Construction costume / development costume
The bees bring in this tracht until the middle/end of April. At this time the bees come out of the winter. Mostly they are overwintered with only one frame, i.e. only one layer of combs. At the beginning of the build-up harvest, pollen is the main protein source needed to feed the queen and brood. Around the end of March, the winter stores of honey and winter food are used up and the bees have to start collecting nectar. At this time at the latest, a second frame is added to give the bees more space for development.

The build-up harvest is completed with the beginning of the fruit blossom, but can last correspondingly longer after a long winter. The plants that are approached by bees during this period are for the collection of pollen: Birch, alder, hazelnut, crocus and other plants that bloom early in the year. Nectar sources include blackthorn, Norway maple, sycamore, field maple and horse chestnut. But other spring flowers and trees that bloom early in the year also provide food for the bees.

Early harvest

In the period from mid-April to the end of May, the build-up harvest turns into the early harvest. Now the development of the colony is so far advanced and the hive is so abundant that the first honey stores are created. It is therefore time to install a honey chamber in your hive.

What your bees collect as early forage again varies from region to region. Cornelian cherry, blackthorn, garden and meadow flowers, dandelion, clover and rapeseed are the dominant sources of pollen. But fruit trees, berry bushes, such as currants and strawberries, are also part of the early harvest, depending on the region.

In spring, things usually happen explosively fast and it seems as if nature unfolds its flowers everywhere "from one day to the next". A well-developed and healthy colony of bees now finds plenty of food everywhere, and the first mass crops are formed. Mass harvest means such an abundant supply of food that the bees build up large stores of honey, which the beekeeper can then harvest as honey. Early harvest honey tastes aromatic to mild and varies in consistency and color and aroma depending on the harvest.

During times when one type of grape is predominant in your region, varietal honeys may result. For example, if your bees live in a region where a lot of canola is grown, they will most likely use canola almost exclusively as a source of pollen and you will be able to obtain a varietal canola honey.

Early summer costume

The next stage of the costume is the early summer costume, which lasts from May to about mid-June. Here, the range of tracht changes and becomes much more varied. There is now a wide range of meadow herbs, ornamental plants grown in gardens or parks, and useful plants. Not only in the countryside, but especially in the city, there is a wide range of food sources for bees. For example, maple varieties, black locust, hawthorn, clover species and berry fruits.

In the early summer bees can also use the first honeydew. Especially on horse chestnut and maple aphids may have spread, which provide the bees with the honeydew. The spruce whisk scale aphid can settle on spruce as early as the end of May and provides plenty of honeydew.

Summer costume

Summer honeydew extends from mid-June to the end of July and sometimes to mid-August. During this time, you can call your honey summer honey, Sommertracht or Sommertracht honey. Depending on the region, exemplary sources of honey are: clover, blackberries, raspberries, sweet chestnuts, wild herbs and all kinds of ornamental and useful plants in German gardens and parks.

Regionally, varietal honeys may be produced, for example, during the linden blossom, in the growing area of sunflowers or during the asparagus season. Pay attention to whether your bees - if you are in the growing area of asparagus - bring in red pollen. These come from asparagus (Asparagus officinalis).

Now is also the time when there can be a hive gap, a time when your bees cannot find enough food. This is most common in areas with a high monoculture of agriculture. When canola is grown as the predominant flowering crop, there is a lack of masses of tracht once it is harvested. Now the bees have to fall back on other sources of pollen, which are often lacking because modern agriculture uses weedkillers and the grass in fallow meadows is too often kept so short that nothing can bloom.

Spättracht

After the summer harvest, at the end of July or in mid-August, the period of late harvest begins and lasts until the end of September and in warm late summers even until October. Late harvest is everything that your bees bring in after the second honey harvest, i.e. after the early harvest and after the summer harvest.

With a few exceptions, this harvest is usually not usable for honey harvesting, as it remains rather small in quantity and should then serve the bees as winter stock. In areas where sunflowers or buckwheat are increasingly cultivated, however, the late harvest can be abundant enough so that a further honey harvest makes sense. Late harvesting is also possible in heathland, where everything does not begin to bloom until late summer. Here now bloom broom heather and bell heather species.

Think about harvesting the buckwheat crop only if

you live in a region where buckwheat is grown throughout the country. Buckwheat flowers from the end of July to the end of September and provides abundant nectar. In suitable regions a varietal buckwheat honey harvest is possible.

Mild fall weather can be a problem for you as a beekeeper in regions where white mustard and oil radish are used as green manures in agriculture. If the weather is very mild, there will be an autumn bloom of these two plants, which can be used by your bees to bring in honey. This makes the bee colony more vulnerable to the varroa mite, as winter dormancy should actually begin at this time.

Honeydew

Honeydew introduction is possible from the end of May to the end of September, but occurs only irregularly, as the presence of aphids cannot be predicted. As a result, honeys that can be called varietal honeydew honey are rare and more expensive. Host plants typically infested by aphids are: Pine, linden, spruce and maple. Honeydew honey from these sources of the trade - if it is pure in variety - is often referred to as forest trade or forest honey, even if the trees were located in the city, for example. In the region of the Black Forest, bees also like to collect honeydew from silver fir, which produces pure fir honey there.

It is usually the case that nectar and honeydew are collected together, so that the resulting honey is a mixture. However, pure honeys, for example from lime trees and sweet chestnuts, are also possible as a mixture of nectar and honeydew.

JOB ALLOCATION IN THE BEE STATE

To ensure that the bee colony - also known as the bee - functions smoothly and that the 10,000 to 40,000 individual bees all know what to do, various tasks have to be performed. First of all, there is the queen, who lays the eggs and takes care of the offspring. In addition, there are a number of other tasks for the workers.

Exactly which job they do is linked to their age. In the first twenty-one days after hatching, the bees go through the following schedule: cleaning, brood care, comb building, honey preparation, guard duty. After that they become collectors. However, depending on the area in which there is an additional demand, the honey bees can change their job area and be flexibly deployed where they are needed.

Depending on the performance of her duty, the worker bee has active glands. Feed glands for the nurse bees, wax glands for the builder bees and venom glands for the gatherers. The activation and inactivation of the various glands occurs through the hormones of the bee in the course of its life development. With increasing "life experience" jobs become more and more demanding until the older bees become collectors.

Cleaning bees

A cleaning bee becomes a bee immediately after hatching. In the first days of life, the cleaning bees scurry around the brood nest area and clean the used combs from the remains of the previous brood. They prepare the combs for brooding, i.e. the insertion of an egg by the queen. To do this, they use their mandibles to remove the larva's feces and the remains of the cocoon. Finally, they line the comb with a disinfectant coating of propolis.

They are also responsible for temperature regulation in the hive and can generate heat by rhythmically tensing and relaxing their flight muscles. Cleaning bees are furry, by which young bees can be easily recognized. As they age, the bees lose more and more bristles.

Nurse bees

The nurse bees are bees between four and ten days old. They take care exclusively of the brood care and the queen. In order to feed the larvae, nurse bees have active head glands in which a secretion, the feeding juice or royal jelly, is formed. This secretion is used to feed the youngest larvae up to the age of three days. Furthermore, nurse bees ensure that pollen and finished honey are ready and fed from day 4 to those larvae that are to develop into workers or drones. The larva of a new queen is fed the high-quality royal jelly instead of the mixture of pollen and nectar throughout its development.

A certain number of nurse bees form the queen's

court. They are always around the queen and feed her with royal jelly to keep her strong. The difference between the normal foraging mash and royal jelly is that the latter has a high proportion of mandibular gland secretion added. The normal feed mash instead comes from the hypopharyngeal glands and only a very small percentage is added from the mandibular glands.

Construction bees

The builder bees are characterized by the active wax glands on the underside of the abdomen. Older bees can become builder bees again in times of greater need, mainly in spring after swarming when a new hive is occupied. In this case, the inactive wax glands reactivate.

If no complete new construction is pending, there are only relatively few construction bees that take care of the maintenance of the combs. The wax platelets from the fat body are kneaded between the mandibles, with the addition of some mandibular gland secretion, and thus made malleable. In addition to repairing honeycomb defects, builder bees are also responsible for capping brood and honey cells.

When building a new honeycomb or doing a major repair, builder bees work together as a team. They form a net of bees that produce the wax and then pass it on to the builder bees below the net, which shape it and grow it. This has the advantage that the temperature is high enough, at 30 to 40 degrees, that the wax remains pliable and easy to work with.

The builder bees have a greater significance for the colony composition than is apparent at first glance. The queen scans the empty combs with her antennae and, depending on the size of the comb, lays a fertilized egg for a worker bee or an unfertilized egg for a drone. The comb size for the workers is 5.2 to 5.4 millimeters in diameter and 10 to 12 millimeters in depth. The combs for the drones are 6.2 to 6.4 millimeters in diameter and 16 millimeters in depth.

The freshly produced wax has a pure white color. Through contact with pollen, honey and of course the brood, it gradually darkens and takes on the typical yellow color.

Honey makers

Nurse bees, which do not become builder bees, become honey makers directly. These are the bees that are responsible for managing the honey supply. They are also responsible for transferring the honey.

It is important that the larvae of the bee colony always have enough honey to grow up healthy and become strong workers or drones. That is why bees like to store their honey near the brood nest, the core of the bee colony. This area is called the foraging ring. The queen sits there together with her nurse bees. When food is plentiful, the worker bees carry the honey around. That is, they pick it up from the combs near the brood nest and carry it to clean honeycombs further away. There it is stored capped to build up a reserve.

However, not only the transfer of already finished honey is part of their job, but also the storage of freshly brought in tracht. Through trophallaxis, i.e. social feeding, the collectors transfer the contents of their honey bladders to the honey makers, who distribute the already predigested mix in the honeycombs. The honey makers also take tree resin, which is used to produce propolis, from the collectors and stow it in the hive. So does pollen. Pollen that is not used directly to feed the larvae is then stored by the honey makers in the outer combs of the feeding wreath and mixed with glandular secretions and unripe honey. This creates the so-called "pollen bread" or "bee bread", which can then be used as food when needed.

In addition to their honey managing job, honey makers are also responsible for carrying garbage and debris out of the hive. The job description is generally no longer set in stone from this age onwards and so it may well be that you see workers in your hive running over the honeycombs without being able to say exactly what job they are doing.

Guardian

From about the 18th to the 20th day of life, bees work as guard bees. This also represents the transition from the hive bee, which only stays in the hive, to the flight bee. Guard bees stay in the entrance area and make short surveillance flights near the hive. Every bee that wants to enter the hive must be checked by the guard bees. During this process, the bees make contact through the antennae and the guard bee checks the arriving bee for hive odor. Only those that have the correct odor and can thus be identified as belonging to the hive are allowed through. All others are attacked with the poison sting.

Production of venom begins on day 3 of life as an adult bee and ends with a full venom blister on about day 15. In summer bees, those incubated between March and August, venom production ends on day 20 of life.

Not only other bees and insects, such as hornets and wasps, are repelled by the guard bees. Smaller vertebrates, such as birds and shrews, can also enter the hive as predators. Especially the larger enemies cannot overpower the relatively few guard bees alone. Then the alarm pheromone is released and other workers rush to the aid of the guard bees. Together the intruder is killed and then, if it is impossible to drag it out of the hive, it is literally mummified with propolis. In certain situations, the guard bees also tolerate drones and workers from other colonies. Stray workers or bees that have lost their own colony can settle in a foreign colony and are taken in. However, when the honey supply is scarce, colonies may rob each other and steal the honey supply from the foreign colony. Guard bees react more aggressively to foreign bees when these robberies occur more frequently and no longer allow them to pass.

Collecting bees

Older worker bees, from about three weeks of age, become forager bees. They now belong to the flying bees and leave the protective nest to go about their work: Collecting honey, honeydew and pollen to provide food for the colony. Once a forager bee, always a forager bee. As a forager bee, it is their job to watch the forager bees and thus be directed to the best source of honey. They collect nectar, honeydew, pollen, water and, if needed, raw materials for propolis production.

The behavior of the collecting bees is commonly referred to as "bloomstet". This means the following: As you have already learned, forager bees fly out to find particularly lucrative sources of honey. They return to the hive and perform the tail dance. The bee that advertises the most promising tract is followed by the forager bees. They fly to this tracht and also perform the tail dance on their return, so that gradually all the collecting bees are induced to fly to exactly this tracht source. This behavior is called blütenstet.

Whether a tracht is considered lucrative or not depends on the number of flowers and the sugar content and general amount of nectar. The forager bee flies out over and over again to bring in nectar and pollen until it dies on its last flight. After two to three weeks, the forager bees clearly show their exertions. Usually the wing tips are torn and the light-colored banding on the abdomen disappears more and more as the bristles break off. The abdomen becomes darker and darker and the bee eventually loses the ability to fly and dies outside the hive.

Track bee

Being a tracker bee is the most dangerous job in the bee colony and is reserved for only a few foragers. They are scouts, explorers of the world and have probably the most important task in the hive: finding new sources of honey. In this way, they ensure the survival of the colony. During swarming, the tracker bees have the job of finding a new home and guiding the swarm safely there. Within a few moments, the tracker bees manage to form an error-free memory of the location and, after the reconnaissance flight, fly back to the swarm and not back to the old hive.

If you see the first bees flying out of your hive early in the morning or after the rain, these are the tracker bees that fly out and go in search of grape. They fly in a maximum radius of 3 to 5 kilometers around the hive. They are constantly searching for nectar and honeydew. In the spring, they also look for sources of water and pollen to meet the extra demand for liquid and protein.

As you have already learned in the chapter on the physiology of bees, the trace bees can mark found sources of tracht with scents. In addition, a sample of the tracht is brought home in the honey bladder for the foragers to taste. The tail dance, scent marking, and smell and taste of the new tracht source then show the foragers a safe path to the new food source.

Winter bees

The winter bees are those bees that are hatched from the brood in the fall. Their task is to take care of the queen in winter and keep her healthy. In terms of anatomy and physiology, winter bees do not differ from summer bees, only their task is different. They no longer fly out to collect grapes, but stay exclusively in the hive. Their primary job is to provide heat production by forming the winter cluster. The queen continues to be supplied with food by her court.

Once the winter is over and the breeding season slowly picks up, the winter bees have to adapt flexibly to their new areas of responsibility, becoming cleaning bees, nurse bees, builder bees, guards, collectors and tracker bees. Thus, due to the effort of work, they begin to age like the summer bee and die at the end of March to the beginning of April. At this time, the first generation of summer bees has taken over the work in the hive.

The queen
In the technical language of beekeeping the queen is called "queen bee". The term "hive mother" is also commonly used to refer to the queen. She is the only bee that survives for several years. You have already learned a few things about the queen, including that she emits queen substance, a pheromone mixture that signals to all bees that they have a strong and healthy queen. Her biggest and most important task is reproduction, egg-laying. She performs this work year after year from March to August. The process is called "setting pins." Each day, she lays up to 1,200 eggs - called pins - in the prepared honeycomb cells. Converted, she produces and lays 80% of her own

weight in eggs every day. She can only maintain this performance with the protein-rich royal jelly as feed juice.

Two situations can lead to the raising of a new queen. The old queen has reached the end of her life and dies shortly before or shortly after the hatching of her successor. In this case there is no division of the colony, it is also called silent transfer. In the other case, the colony has grown so large that it divides. In this case, the old queen leaves the hive with part of the colony shortly before the new queen hatches.

The entire development of a queen lasts 16 days. 10 days she spends in the egg and as a larva. 6 days lasts the period of pupal dormancy, the metamorphosis.

Drone

The male bee - born for one purpose only: to mate with the queen of a foreign colony.

Drones are produced by the bee colony only when they are needed, which is at the time of swarming. By mating only queens of a foreign strain, drones ensure the exchange of genetic material and prevent inbreeding.

Drones are created from unfertilized eggs. Genetically, therefore, they have no father, but after successful mating they can become the father of a completely new bee colony.

In the first days of life, the drones are fed with forage sap by the workers, after which they provide for themselves from the colony's honey stocks. They need large amounts of protein, i.e. pollen, for sperm production. Sexual maturity is reached between the 8th and 12th day of life. By then, drones have produced between 8 and 11 million sperm.

The drone makes a few short orientation flights in the first days of life and flies out to drone collection sites, where male bees are always on the lookout for a suitable queen. Once it has found one, it dies during mating after transferring its sperm to the queen.

An unsuccessful drone can live 30 to 40 days and is mostly driven out of the hive by the workers before overwintering. This is called a drone battle. Rarely, however, individual drones may be tolerated in the hive during the winter and may overwinter with them.

THE NEST

You can consider the brood nest of your bees as the heart of the hive. This is where reproduction takes place, which is why this area deserves special attention. In a smaller colony the brood nest extends on one frame, a larger colony, also called economic colony, usually has a brood nest distributed over two frames.

Use warm days for viewing the brood nest of your bees. The air temperature should be above 20 degrees so that the larvae do not cool down. Record all your observations in the hive chart. This is your record keeping of the hive. For each hive you have a separate hive card, in which you note everything that happens around your bees, and also what work you have done in connection with the bees.

Here is also space to enter what observations you have made in connection with the brood nest. In the center of the brood comb you can see the brood cells, which are bordered on the outside with pollen-filled comb cells. The whole is bordered with cells full of honey, which is either already ripened and ready for feeding or still fresh and unripe just brought in. This characteristic structure of the brood nest is called a feeding ring.

The metamorphosis of the bee

In the brood cells you can find different things. An egg, called a pin, a round maggot, an elongated maggot, or the pupa, which undergoes holometabolous metamorphosis and changes from the much simpler-built larva into the adult insect (imago).

This development - from egg to imago - is astonishingly fast in bees: within just three weeks, the finished bee hatches from its pupa. Other insects have development times of several months, some even of several years, such as the cockchafer or dragonfly species. To make this rapid growth possible, nurse bees do their best to supply the brood with energy-rich food juice.

The fertilized egg forms the first of three growth phases of the bee. As soon as it has been laid by the queen, embryonic development begins inside it up to the larva, which hatches from the egg on the 3rd to 4th day after oviposition.

Now the second phase of development begins: the growth phase. Here the larva is supplied with food by the nurse bees and grows quite quickly into the finished round maggot, which reaches its final weight of 150 to 160 milligrams within just five days. Up to this point, the bee maggot has shed its skin four times, shedding the old exoskeleton, which has become too small, to obtain a new, larger shell. Remnants of this old exoskeleton can hardly be found, since it consists of proteins and is eaten directly by the maggots so as not to waste valuable proteins.

A round maggot that has reached its final weight becomes an elongated maggot. This means that it no longer lies curled up in its brood cell, but stretches out lengthwise. The workers now cover the brood cell and the third phase begins: differentiation or metamorphosis. The stretch maggot now spins a fine cocoon around itself and becomes a prepupa. Now the larva sheds its skin for the fifth time, this happens on about day 13 of its development. This produces the finished pupa, inside which the larva now acquires a more complex and differentiated form and becomes an imago: the finished, adult bee.

At the end of the pupal dormancy, around day 21, the last molt occurs, the sixth in number. This last molt opens the lid of the brood cell and the finished insect hatches.

THE SWARMING

For you as a beekeeper, it is imperative that you understand the dynamics of swarming to prevent uncontrolled swarming of your colonies. Swarming that you do not control is bad for the colony and for beekeeping for several reasons:

• Especially in urban areas, a feral swarm will have difficulty finding suitable housing and will die if not captured by a beekeeper.

• A swarm of bees flying around freely can easily cause displeasure among your neighbors.

• The breeding successes of the last few years are negated by feral bee colonies, as these basically react more aggressively to defend themselves against predators. A mating of breeding colonies with wild bees that have been released into the wild can lead to the fact that the colonies kept by humans also gain aggressiveness again.

So it is your job as a beekeeper to prevent uncontrolled swarming through good management and sufficient knowledge of the physiology of swarming.

Swarming

There is no "one" reason for your colony to swarm. Instead, it is an interaction of several factors and you can only tell if your colony is in swarming mood if you regularly observe it closely and know its behavior well. Several points must come together for a colony to really be in swarming mood. First of all, the development and the building up of the brood must be completed. A colony must first come out of the winter well before it can even think about swarming. It must be a healthy colony, an ailing colony does not have the strength to prepare for swarming. Swarming also takes place only in a time when there is enough honey. The period for this is, depending on the weather, between May and mid-July.

A colony will swarm more easily if there is a lack of space. If the magazine is full and space becomes scarce, division of the colony by swarming is prepared. Another reason to divide a strongly grown colony is that the queen substance in the colony is too diluted. If the concentration of the pheromone mixture drops too much, the workers begin to prepare swarming.

To do this, they create queen cups. These round cells are usually placed at the edge of the brood comb, often they extend onto the wooden frame, which makes it easier for you as a beekeeper to recognize the royal cups. As soon as the queen has pollinated one of these queen cups, i.e. has laid a fertilized egg in it, the workers develop it into a queen cell and the rearing of a new queen begins.

While the new queen is being grown - usually several potential queens develop at the same time - the old queen must be made fit to fly. However, the active ovaries add far too much weight, so that the queen would not be able to fly in this condition. Therefore, she is put on a diet by her court and gets only an economy portion of food.

At the same time, she is "forced to exercise" by workers touching the queen's thorax with their legs and generating vibrations. Thus, the queen has to walk over the combs as fitness training. Through this sports program, the queen loses about 25% of her own weight and will also not lay any new eggs during this time.

At the same time, hormonally controlled processes begin to reactivate the wax glands in the collectors and other workers, so that the swarm has many active builder bees that can then start building new combs directly in the new housing. Shortly before the final start, those bees that "leave" the hive with the swarm fill their honey bladders with enough provisions to survive the journey and the construction of the new hive. The new swarm takes about 500 grams of honey with it in this way.

The extract

Now all preparations are made for swarming. Shortly before it finally starts, calm returns. Often the bees hang as a swarm cluster in front of the flight hole, so that you can recognize them very well. At this point, at the latest, you should have made all preparations to successfully capture the swarm, because you can no longer stop the swarming at this point. If the start signal is then given by the tracker bees, the swarm is ready to go.

So it all depends on the track bees. These determine the final time for the new swarm to leave the nest. To determine this time, the tracker bees commute from indoors to outdoors. They record the weather conditions, which must be warm and dry for successful swarming, and at the same time check the status of the queen cells.

When they are ready for takeoff, generate a vibration on the honeycombs in the frequency of 200 - 250 Hz. This frequency is absolutely specific for swarming. More and more tracer bees of the swarm join and thus alert the waiting bees. The temperature in the hive rises simultaneously with the tension and finally (literally) all the involved trace bees run off, dragging the workers and the queen with them - the swarm, consisting of two thirds of the adult bees of the hive, rises.

However, only for a short flight, which is very convenient for you as a beekeeper. The swarm settles at a nearby, usually elevated point, such as the branch of a tree, and forms the swarm cluster there. The bees literally cling to each other and wait. The length of time the swarm spends at this point is variable from a few hours to a few days. It depends entirely on how long it takes for the tracker bees to find a new suitable dwelling. During this time the swarm is very vulnerable. If the good weather suddenly breaks, for example if there is a summer thunderstorm, the entire swarm may die.

The search radius of the tracker bees is about 2 to 3 kilometers around the formed swarm cluster. Once a tracker bee has found a possible dwelling, it returns and performs its tail dance. Sometimes some bees break away and explore the advertised dwelling as well. Eventually, the swarm as a whole decides on the best advertised dwelling. Once the decision has been made, the tracker bees activate the other swarm bees by vibration and nudging, and the swarm sets off as a whole to move into the new dwelling.

Formation of a new people

The bee swarm only becomes a bee colony again when the housing is occupied, the combs are built, the first eggs are laid and the first crop is brought in. In the first place is the dwelling. The bees prefer living cavities with a volume of about 40 liters. Preferably at a height of about 5 to 6 meters, with a small entrance hole facing south or southwest. Of course, the dwelling should be dry inside and not allow drafts. Smaller cracks and holes the bees like to putty with propolis to keep out all drafts.

Once they have moved into their new home, the workers immediately start building honeycombs and feasting on the provisions they took with them from the old burrow. Depending on when a swarm has fledged, it now has more or less time to bring in the first crop to ensure survival. If the weather turns bad too quickly, or if a hive gap opens up for other reasons, the food supply can quickly become scarce and the swarm starves to death. As a rule, swarms that fly earlier in the year have a higher chance of survival than swarms that do not swarm until late July or August, for example.

A few days after moving in, the queen will begin laying eggs again once the workers have created brood combs and there is a sufficient food supply to provide the larvae with foraging sap.

What happens in the old people?

The first swarm to leave the colony is often called the preswarm. Depending on the strength of the remaining old colony, a second, much smaller swarm may form. This is then called a post-swarm.

As you already know, several queen cells were created shortly before swarming, in which the next queens develop. In most cases, the forerunners leave before the new queen hatches from her cell. The workers bridge this time by continuing to do their jobs.

If the first young queen is then ready to hatch, she announces herself with a certain vibration pattern from inside her queen bee cell. This is the question whether the old queen has already moved out of the colony. As a beekeeper, you can even hear this sound, which is reminiscent of a croak, from outside the hive. If the old queen now responds, also by a vibration pattern that she generates with her flight muscles, the new queen knows that she still has to wait with the hatching and remains in her queen bee cell. From time to time she asks the question again by vibrating and hatches as soon as the old queen no longer answers her - when she has left the hive.

Since several young queens will now hatch almost simultaneously, the next queen will soon inquire through the vibration signal. Either, in the case of a very strong colony, a second swarm, the post-swarm, will now form and also leave the hive, or the new queen will kill her competitor still in the brood cell by a sting. If several queens hatch at the same time, there will be a fight among them.

It is the workers, not the new queen, who decide whether a secondary swarm will develop. It depends solely on whether the brood situation and the honey stocks are still sufficient to divide the colony. If this is the case, the workers protect the queens in the queen cells so that the other queen cannot kill them. The post-swarm develops. If the stocks are not sufficient, the workers do not protect the queens in the queen cells and the young queen can eliminate her rivals.

How to start?

In this chapter, you will get a brief overview of all the important issues you need to consider and know about before you acquire a bee colony. Use this section as a kind of guide or to-do list to check that you have thought of everything important and made all the preparations.

Keep in mind that with beekeeping you are taking responsibility for living creatures. Bees, in their own way, are just as high-maintenance at certain times of the year and demand as much time from you as any other pet. So you need to be aware that you will be tied to your bees, especially in the spring and summer. Can you balance this with your vacation schedule, job and family? This is the first question you should ask yourself.

Take enough time and think about the following questions to find out for yourself if you are ready to start keeping bees:

• Are you related to nature, its flora and fauna?

• Are you a calm mind and do you always act thoughtfully and not frantically? Hustle and bustle can flush out your bees and cause them to see you as a danger.

• Do you have some organizational skills and some manual dexterity to handle the bee magazine?

• Are you willing and do you have the time to spend

three to four hours a week with your bees? You will have to spend this time mainly in spring and summer. In autumn and winter correspondingly less.

• If you are allergic to bee venom or have other health limitations that prevent you from performing the physical work on the hive, you should decide against keeping your own bees to protect your own health.

BEFORE THE START

Managing a bee colony profitably requires a high level of expertise that you cannot gain just by reading books and guidebooks. Nothing replaces the tips and tricks that an experienced beekeeper can give you. Therefore, it is worthwhile if you become a member of a beekeepers' association at the beginning of your career and look for a so-called beekeeper mentor. This person can help you with your own colony or show you certain hand movements on his colony. There you can first familiarize yourself with the bees at the beginning. You can look at all the things that you have only read about in theory up close, listen to them and also feel them.

The beekeepers' association in your area can also help you out with valuable addresses and tips. Here you may even be able to borrow equipment or at least get addresses and contact points where you can get everything you need. You can also get in touch with bee experts, specially trained beekeepers, through your association. Finally, the association also offers you the possibility to

get a beekeeper mentor. This experienced beekeeper ac-
companies you with advice and action in the first two to
three years of your beekeeping and is always there for
you and your bees.

WHEN IS THE BEST TIME TO START?

Since bees are seasonally active animals, the starting time
is quite fixed. Depending on how you acquire your first
bees, the right time will also be given.

The easiest way for you to do this is to buy one or
more scions from an experienced beekeeper. The beekeeper creates scions by dividing and thus reducing the size
of a strong colony. The advantage of a scion is that the
colonies grow slowly and you do not run the risk of soon
having a homeless swarm in the garden. However, you
will not be able to harvest honey from a small offshoot in
your first year. The colony grows slowly and builds up
stocks to get well through the winter. Then in your second year, your colony will have reached the size of a
commercial colony and you can look forward to a bounti-
ful honey harvest. Of course, you can also get to your first
colony of bees by capturing a swarm - but you should
really only do this in cooperation with an experienced
beekeeper, otherwise you endanger yourself and the
swarm.

An experienced beekeeper, or if you have very close
support from a very experienced beekeeper, can also

purchase a commercial colony directly in April. These strong colonies have the advantage of providing a honey harvest in the first year after purchase. However, you must be familiar with the measures you need to take to prevent swarming so that your colony does not soon say goodbye to itself.

Again and again peoples are also offered in a discount. This has the advantage that you can often also take over utensils and equipment and do not have to acquire them new. Here the season plays a special role, whether you can move the peoples still in the same year or whether it is smarter to let the peoples overwinter at the usual place. To take over an estate is therefore also only recommended for experienced beekeepers.

PREPARATION

Before your bees can move in with you, you should have acquired various things. This includes the housing for your bees, but also your personal protective equipment.

The hives

Hive is the technical term for the housing of your bees. There are regional differences in beekeeping forms, of which magazine beekeeping is the most common. Other housing forms are: Stülper, beehive and beehive box. In the case of the beehive, honey harvesting is not possible, this form is suitable only for keeping bees. The magazine hive is a so-called vertical hive.

The individual hives have a bottom on which you can place as many frames as you like, almost like a construction kit, which you can extend upwards as you like. Frames are wooden frames in which you hang the frames that the bees use for honeycombing. On the top frame there is a lid that closes the whole structure. An advantage of the magazine hive is that you can stack as many frames on top of each other as you like and thus, depending on the size of your colony, create a better space or also reduce the space, which is useful in the fall, for example. The hives can be made of wood or styrofoam. Wooden hives are classically used, as these are also more environmentally friendly than Styrofoam variants in terms of ecological waste recycling.

Inside the frames hang the already mentioned wooden frames, which beekeepers also like to call frames. The frames are the space that the bees fill with honeycombs. To stabilize the combs and make it easier for the bees, the frames are often braced with fine wires or even with a middle wall, which you as a beekeeper can specify so that the bees can place their combs on both sides of this middle wall. In this way, you make it easier for the bees to build, since you are already giving them some of the wax.

Only through the removable frames a honey harvest becomes possible, because the frames can be spun.

The location

Protected from the wind and stable are the indispensable requirements for the location of your hive. It was already mentioned above that bees prefer a higher location, so be careful not to place your colony in a depression and never place the hives directly on the ground. Work here, for example, with pallets or other pedestals to place the hives higher.

The flight holes should ideally face south or southwest, and the area directly in front of the flight hole should be as free and little used as possible. Here, a free area of about three meters in diameter has proven successful. There are recommendations, how much place your property or the location of the bees, approximately should exhibit: per bee colony 100 square meters are suitable. This is of course - especially in the city - often not complied with. Think here therefore of your neighbors! In order to prevent discontent, it is advisable to coordinate the location with the respective neighbors. Be open to your neighbors' concerns, fears for their children or possible allergies to bee venom. If there are reservations that cannot be dispelled, you should look for another location and possibly consider an allotment garden.

In some areas, so-called protected areas or restricted areas, beekeeping is not allowed. Inform yourself in advance at the responsible authorities to avoid warnings or strict conditions.

BEE TRANSPORT

Transporting bees basically takes place in the hive. This is a very delicate process, which without expertise can lead to the death of the entire colony. Therefore, pay attention to essential things during transport and, in case of doubt, get an experienced beekeeper to help you.

Death during transport occurs due to the so-called scorching. This happens when the colony overheats inside the hive. If the temperature rises, the bees increase their activity and beat their wings to cool the colony. However, you have closed the flight hole in advance so that, of course, no bee is lost to you during the journey or even escapes in your own car and attacks you.

Now no cool air can enter the hive because the flight hole is closed. The activity of the bees therefore leads to the opposite: it gets warmer and warmer. This is fatal for the wax combs, which lose stability with increasing temperature and eventually break through. Ripe and unripe honey drips down uncontrollably, causing the sensitive insect bodies to stick together. This is usually the death sentence for the entire colony.

So transport your bees only on sufficiently cool days. To prepare for transport, carefully close the flight hole with foam. In order not to exclude any bee from the hive and forget it at the old location, you should do this only outside the flood times: i.e. early in the morning, late in the evening or on a rainy day. Make sure the hives are stowed very securely in the car, trailer or truck bed with tie-down straps and other measures so that nothing can slide or tip over. After you have arrived at the desired location and the hive is satisfactorily in place, remove the foam from the flight hole and move away quickly. Bees do not like transportation and can react accordingly annoyed and aggressive.

Give your bees a few days to arrive and familiarize themselves with the new environment. Use this time as well: observe and get to know your bees. Especially in these first days you can observe a lot of trace bees and bees stinging at the flight hole to spread the hive odor. After a few days you can carefully open the hive for the first time and also look at your bee from the inside. What can you see? Feel free to take advantage of your beekeeper mentor's knowledge and ask him or her to explain what you see.

Requesting a health certificate - the necessary paperwork

Please note that a health certificate is required for moving a colony of bees across county borders. This is specified in the Bee Disease Ordinance and serves the purpose of ensuring that only healthy colonies are moved. This prevents the spread of the bee disease "American foulbrood".

In addition, you must notify the veterinary office responsible for your district of your beekeeping and present a copy of this health certificate. In some federal states, in addition to this notification, you must also register your keeping with the responsible animal disease insurance fund. The beekeepers' association responsible for you can tell you exactly what you have to do and where you have to register.

It also makes sense to take out liability insurance for the bees. Often the insurance is included in the membership of the beekeepers' association. If this is not the case, the people in charge there can tell you exactly what to do.

The beekeeper

There are around 600,000 bee colonies in Germany. The average beekeeper is male and 57 years old. At least the latter is changing in these years. Beekeeping was long considered a hobby for "grandpas and old people," which thankfully is no longer the case these days. More and more young people and even women are finding enjoyment in beekeeping and keeping bees. But what is it that makes a beekeeper - and what does he need to successfully pursue his hobby?

Until the 18th and 19th centuries, wild bee colonies were exploited by humans. Here the combs were simply cut out, the survival of the colony was secondary. Today, the value of each individual bee colony is widely known. The pollination service that bees provide is desirable and thus the survival and health of each colony is increasingly the focus of beekeeping. Honey is harvested only when it does not harm the bees.

WHAT DO I HAVE TO DO AS A BEEKEEPER?

Wild colonies exhibit two behaviors that a beekeeper wants to avoid in his colony.

1. Swarming occurs regularly as the colony steadily increases in size.

2. After a few years, the current dwelling is completely abandoned and a new, clean home is sought to prevent heavy soiling and parasite infestation.

As a beekeeper, it is your job to stop this behavior. Accordingly, you take responsibility for keeping the hive clean and effectively controlling diseases.

Brood chamber and honey chamber

In magazine beekeeping, it is possible for you to spatially separate the brood chamber from an area that is used exclusively for honey storage: the honey chamber. You achieve this by inserting a grid. This grid is so fine that the queen cannot pass through it. The workers can still pass through and use the honeycombs created in the honey chamber solely for honey storage.

If your colony grows strongly in spring, add another brood box. This will give your colony room to grow and store more food. Swarming is prevented by this and by the honey harvest.

Controlled propagation

A people does not live forever. At some point it dies, no matter how good your management is. So, in order to preserve the honey bee as a species, it is essential that you take care of the reproduction of your colony. To prevent uncontrolled swarming, form offshoots from your own colony.

For this you need an empty hive. Remove several brood combs from the existing colony, add empty combs and frames with center walls on which the builder bees can create new combs. Place this construction in the new hive. Bees attached to the removed combs are transferred and form the colony's offshoot.

Promote hygiene: Honeycomb hygiene

Pathogens collect in the wax of old combs. These are the reason why the bees leave their old hive after a certain time and settle again. Since you prevent this as a beekeeper, it is necessary to remove old combs from the hive from time to time. In this way, you stimulate the construction of new combs in your bees, which again have a better status hygienically.

Feeding

In rare cases, there are hobby keepers of bees who do not have the goal of extracting honey. However, most beekeepers want to extract honey from their bees and use it commercially or for their own use. By taking the honey, you are depriving your colony of its livelihood: food stores. So it is your responsibility to provide winter feedings and emergency feedings when there are gaps in the hive to keep your bees full and healthy.

HOBBY - OR PROFESSION?

There are some terms that differ depending on whether you are beekeeping as a hobby or commercially. If you are a pure hobby beekeeper, you are keeping bees in your own backyard. You do not pursue commercial purposes with it, you do not have to register a business.

One speaks of an apiary when keeping bees commercially. You are then called a professional beekeeper or part-time beekeeper and your goal is to earn money with your bees, possibly even full-time. It is not enough to own only five bee colonies. On average, professional beekeepers have between 50 and 500 bee colonies.

As a beekeeper, you devote yourself entirely to the breeding of the honey bee. You do not extract honey, but are purely responsible for reproduction.

There is a recognized vocational training program for beekeepers that lasts three years and ends with a journeyman's examination. The official designation for this is "Tierwirt, Fachrichtung Beekeeping".

THE BEEKEEPING EQUIPMENT

The equipment necessary for beekeeping and apiculture can be roughly divided into two tasks: the equipment you need for beekeeping and handling your colony, and the equipment you need for honey extraction and processing.

This book introduces you to the equipment for magazine beekeeping. This form of beekeeping is the most common in Germany today. It is a mobile hive. In mobile hives, the combs are built on the frames, which you can hang in the hive with the frames and also take out again. In this form of beekeeping, the honey can be extracted by spinning and contains hardly any wax. In contrast to this, the honeycombs are firmly installed in the hive by the bees in the so-called stable hive. Here you cannot simply take the combs out. They must be cut out for the honey production. Now you chop the honeycombs and extract the honey by pressing and draining. Accordingly, the honey obtained in this way bears the name: Press honey or drip honey. In these forms of honey is correspondingly more wax content, which significantly changes the taste.

Stable hives have another clear difference and at the same time a serious disadvantage compared to mobile hives, which makes them unsuitable especially for inexperienced beekeepers. You have only limited insight into your colony from the outside, so that you recognize diseases and other problems much later than with the mobile hive.

Magazine hive

There are different magazine hives, which is mainly due to the different size of the frames used. Find out in advance what size of frames is common in your area. This will make it easier for you to purchase material later.

Of course, you are free to choose how big you want your frames, you could build them yourself. However, for easier work it is advisable to stick to the standard dimensions.

A frame forms a layer in the magazine hive, which you can fill with 6 to 12 frames, measured according to the dimensions of the frames. Depending on the intended use, there are also special frames. Half frames, for example, are only half as high and offer space for half-high frames. You can use these half frames as a honey room, for example. Then there are, especially for winter feeding, feed frames, which you can fill with liquid feed. The bottom of the magazine hive usually has a grid to ensure good air exchange. Under this grid you can slide a bottom insert for a tight seal.

The frames

The frames are usually braced with wire to make it easier for the bees to build the combs. This has the further advantage that the combs are more stable during spinning and do not break so easily. The frames are hung in the magazine hives. For this purpose, the top beam, the upper bar of the frame, is about one centimeter longer on both sides. These "lugs" make it very easy to see that they are being used and hung correctly.

The dimensions of the frames differ from each other, sometimes significantly, and you should know exactly what size of frames you need before buying. This refers especially to the purchase and sale of scions, as they are purchased with frames.

In the following table you will find four common dimensions compared. The table describes the normal height of the frames. As already mentioned, there are also half-height frames, for example, for the honey chamber.

Dimension	Width in cm	Height in cm	Honeycomb area in cm^2
German Standard Measure	37	22,3	700
Zander measure	42	22	764
Dadant incubator	43,5	28,5	1096
Dadant honey room	43,5	14,5	522

The frame material is mostly spruce or pine, there are also frames made of the harder beech wood. The wire used is stainless steel, which can withstand treatment with formic acid against varroa mites without corroding.

Between the center walls of the frames, i.e. the center of one honeycomb and the center of the next honeycomb, there is ideally a distance of 35 millimeters.

This distance is called the distance between combs. Between the combs is the aisle that the bees need to walk undisturbed on both combs. This distance is called the comb alley. If the comb alley is too wide, you run the risk of your bees creating an additional honeycomb here in the wild. Since this honeycomb is then not attached to a frame, but has direct contact with the side walls of the frames, you can not simply take out the honeycomb, but must cut you out, as with the stable construction.

To maintain the distance, you can now attach wooden blocks or other spacers to the frames, so that you always know exactly what distance is ideal. You can also use so-called Hoffmanns side pieces. These side pieces are used for the side bars of the frames and are widened so that the frames butt directly against each other and the ideal distance is thus specified.

The center wall
You have already learned about the center wall. It is a predetermined wax panel that stabilizes the combs, makes it easier for the bees to build and reduces comb breakage during spinning. You can either make your own center wall if you have already been able to obtain your own wax from your colony, or you can purchase the center walls. Various walls of different dimensions are offered, made either rolled or cast. Some of them are certified pollutant-free.

To connect the center walls to the wire of your fra-mes, lay them flat on the wire and apply a current to heat the wire. In this way, the wax fuses locally with the wires.

If you do not want to use middle walls, it is a natural honeycomb construction. The frames are usually wired anyway, but there is also the variant of making the frames available to the bees completely without wire. This is often done in organic beekeeping. The frames are presented to the bees as so-called empty frames. You also have to present empty frames to your bees for the creation of drone cells in the brood nest.

Wild construction of combs is usually undesirable, because then you can not harvest the honey by centrifuging. In certain forms of beekeeping, such as heath beekeeping, wild construction is deliberately preferred.

However, if you have magazine hives, it is advantageous to prevent game building. You do this by maintaining accurate spacing between the frames. Wild construction of combs can occur in the free area above the last frame, under the roof of the hive. To prevent this, cover the upper frame with an impermeable material. This can be a foil, gauze or the like. Gauze has the advantage that it is permeable to air and any moisture does not accumulate in the hive. With foil, this moisture can lead to mold in case of doubt, which you should avoid at all costs.

The beekeeper suit

Today's honey bees are much more docile than wild bees and allow an experienced beekeeper to work without gloves and face protection. If you are inexperienced in handling or if your bees are currently in a hive gap, it is advisable to always work in full protective gear, as the bees defend their stores vehemently, especially when food is scarce.

The beekeeper suit is airy, but tightly closed. It completely covers the arms and legs and is made of a rough, white fabric. The cuffs ideally have elastics or the like that allow you to vary the width so that the clothing fits snugly and doesn't allow the bees to crawl in.

Hide your hair under a hat or cap with a matching veil around it to protect your face and sensitive neck from the bees. This will also prevent bees from getting caught in your hair and stinging out of sheer panic.

Your feet you pack in half-high, closed work shoes and for the hands there are gloves with long cuffs that fit tightly.

Clothing for honey processing

You handle food at the moment you remove the combs for honey extraction. Your clothes must be accordingly clean and complete. Your hair is to be tamed under a head covering, for example a hair net, your everyday clothes are replaced by overalls or a smock. Always work hygienically.

Smoker

As for all living creatures, a fire is also an alarm signal for bees. They get ready to flee the hive in case the fire approaches. Therefore, the smell of smoke causes the bees to return to the combs and fill the honey bladder.

You can take advantage of this behavior with a smoker to calm restless colonies for a short time and to be able to carry out work. In the past, the smoke was produced with the beekeeper's pipe, also called Dathe pipe. The Dathe pipe takes its name from the beekeeper Dathe, who used it for the first time and described its use. Nowadays, the smoker is used almost exclusively, which produces a larger volume of smoke and is more beneficial to the beekeeper's health than the pipe.

> Be sure to buy a smoker that burns with a diameter of ten centimeters. Experience shows that smaller smokers work more unreliable.

The material to fire your smoker are for example wood chips, hay, smaller branches with dry leaves, herbs, dry wood chips and comparable, easily combustible material. For kindling, use newspaper or egg carton. With fine wood chips or shavings, then add to strengthen the fire, and then pile on the coarser material.

After you have used your smoker, it is good beekeeping practice to responsibly dispose of and extinguish the embers so as not to cause a fire.

Smoker alternative: the atomizer

Water has a similar effect on bees as smoke. Although they do not get ready to flee, they become sluggish and remain sitting on the combs instead of aggressively flying up and attacking you. When using water sprayers, be sure not to chill your bees, so only use the sprayer on really warm days. Honeycombs attract water, so using the nebulizer during honey harvest is not recommended.

Honeycomb

A comb rack is a rectangular box in which you can hang combs taken from the hive to look through them, observe the bees or perform work on the colony. A comb hive gives you the possibility not to put the combs on the ground, which can easily lead to contamination. The most common honeycomb frames are made of aluminum, which you can buy depending on the size of the frames. The box is open at the top and on one long side, has a fixed bottom and three fixed side panels. Always be careful with combs that are outside the hive, so as not to lose the queen by mistake!

Stick chisel

Your universal tool as a beekeeper is the hive chisel. This is a flat chisel, usually made of spring steel, with a right-angled bent end and a blade-like ground edge at the other end.

The hive chisel is useful for loosening adhesions. Your bees will putty and wax the whole structure together. Propolis and wax must therefore be removed with the hive chisel if in doubt. It helps you especially with glued frames or frames. If honeycombs grow wild, you can cut them off with the hive chisel and cut them out.

Bee broom

If you want to remove combs for honey extraction, you must transport the bees on the combs back into the hive. Knocking the bees off the comb is not recommended for one simple reason: honey splashes out, gets lost to you and sticks to your bees. That is why there is the so-called sweeping broom or bee broom. This consists, to guarantee hygienic work, of soft plastic bristles that can not hurt the bees. Over time, you will develop your own technique for sweeping the bees from the combs. Just be careful not to hurt your bees, proceed gently and thoughtfully.

If you want to remove several combs, it is recommended that you do not sweep the bees directly back into the hive, otherwise they will sit down at the next comb and be swept off by you several times - this will cause the bees to become displeased and aggressive. Instead, you can sweep the bees off into a tub or bucket and, after removing all the combs, carefully pour them back into the hive together.

Water

One of the most important utensils for the beekeeper is fresh water. You work with honey and every child knows: honey is sticky. In order to work hygienically, you must regularly clean your utensils and your hands and gloves of the sticky residue. This is also a reason why plastic sweeping brooms are particularly suitable for working with bees.

The bee colony during the year

The changes in the length of daylight are absolutely identical every year. The times follow a fixed pattern that cannot be changed by anything. The weather is not so reliable and depends on many factors. The bee year does not start on the same day or in the same week every year, but flexibly depends on the weather and the onset of spring. This is not only different every year, but also differs from region to region. In more southern regions, spring starts earlier than in more northern, higher-altitude areas. Don't just follow the calendar here, but rely on your eyes and read from nature when your bees become active.

It is recommended for a beekeeper to be familiar with the phenological calendar. Phenology describes phases of growth and development in nature, which are periodically repeated throughout the year. Accordingly, the phenological calendar defines ten seasons, which can be defined by the respective events taking place in nature. Important here are above all so-called phenological indicator plants, on whose characteristic flowering time the phenology is oriented.

EARLY SPRING

The first warm days, often in February or March, herald the beginning of early spring. Now the activity in the hive increases and all corners are prepared for the new breeding season. The bees prepare the brood cells, the queen gets going.

So-called cleaning flights of the bees occur as soon as the air temperature rises to twelve degrees Celsius. To keep the hive clean, the bees do not deposit feces inside. Instead, they collect it in their fecal bladder and now use the first warm days to get rid of the ballast.

The collectors also become active again and go about their job: to provide food for the colony. Early bloomers provide the bee colony with nectar and pollen. Wind-pollinated species, such as hazelnut and alder, now provide a large amount of pollen to meet the colony's protein needs. Flowering willow catkins on native willow species, as well as the flowers of coltsfoot and black alder, provide abundant nectar.

SPRING

As soon as the blackthorn and cornelian cherry in full bloom, spring has arrived. One after another now join the fruit trees: cherry blossom, plum blossom, pear, followed by maple species.

When the apple trees, lilac bushes and horse chestnuts finally blossom, we are in the so-called full spring. Now, at the latest, even the last bee in the hive is actively at work.

The colony begins to grow and to produce its first crop. In order to bring in the first mass crop in a few weeks, a colony must have 30,000 to 35,000 worker bees. During this time, the queen, her court, and the nurse bees are needed to bring the larvae quickly and healthily through metamorphosis. In the hive, the last of the winter bees now mix with the first generation of summer bees.

EARLY SUMMER

Its beginning is indicated by the flowering of native sweet grasses, black locust, black elderberry and hawthorn. Now it is time for the beekeeper to pay attention. If the spring has shown a strong supply of honey, the frames available to the colony are usually filled, and the space in the hive becomes tight. It is now up to you to create enough space for your bees to prevent swarming. But how can you effectively prevent your colony from swarming? You do this by selectively extracting resources: harvesting your first honey! You can also form offshoots at this time if you have a strong, healthy colony. These processes are called "cupping the colony" and are necessary to effectively prevent swarming.

Indicator plants for the beginning of high summer are the sunflower and the summer linden. Bees living in areas heavily used for agriculture now usually have to deal with a hive gap, since the mass harvests in the fields are usually all over. City bees rarely know this kind of hive gap, since there are enough ornamental plants, at least in parks and some front gardens, on which they can help themselves.

In midsummer, pure honeydew harvests may occur. However, this cannot be predicted or planned, since you as a beekeeper cannot influence the development of the aphid population at all.

The colony itself is now slowly preparing for the approaching winter. The brood intensity decreases, there is no more growth. Instead, the colony stores the honey - which is a good time for you as a beekeeper to carry out the second honey harvest. To do this, you remove the honey chamber. Now it is important that you do not put it back on after honey extraction. You are deliberately reducing the size of the hive.

In midsummer, it is very rare to get a varietal honey. Instead, you can look forward to a taste surprise that may vary depending on which flowers your bees have used for foraging. This is different in heath areas, for example.

These areas are also called late harvesting areas. Here, late-flowering plants result in mass harvesting, which can produce a varietally pure honey. Other examples besides the heath are the black forest firs - when infested with aphids. Here a pure honeydew harvest can result. The honey harvest in these areas is still shifted a little later in the year, in order to be able to use the Trachtquellen fully.

LATE SUMMER

Late summer begins quite reliably around the end of August. The apples and plums are now ripe. In the bee colony, all attention is now focused on storing winter stores. No new brood combs are pinned, but all free combs are used for honey storage. The larger the colony, the more stores are needed, so it is a good sign if your colony tends to get smaller towards winter. At this stage you can hardly observe any drones in the colony.

Late summer is the perfect time to control the varroa mite in your colony. The mite has spread almost everywhere in Europe and can hardly be found in a bee colony. Through the spring and summer, it was able to multiply in the hive and is now becoming a health problem.

EARLY FALL

From now on, the bees no longer fly out to collect. They stay in their hive and live exclusively from the stores they have accumulated so far. In order to consume as little as possible, the bees completely stop their brood business.

The bees now living in the hive are the so-called winter bees, which survive until the next spring to ensure the supply of the queen.

In some regions where a lot of oil radish and white mustard is grown, these plants may flower in the fall. If this is the case in your region, it can greatly confuse your bee colonies. Instead of coming to rest and preparing for winter, they will fly out and gather up some tracht.

AUTUMN AND WINTER

Full autumn is indicated by the chestnuts, which now fall ripe from the trees. The quince fruit and walnuts are also now ready for harvest. There is a smooth transition into late autumn, which is characterized above all by the deciduous trees shedding their foliage.

The bees do not fly now. Since there is no prospect of pollination anywhere, flights are a waste of time and energy. On warmer days, the bees fly out briefly at most to "go to the loo," that is, to empty their fecal bladder. Their main task is to keep the queen warm and to heat up the hive. By next spring, a colony will consume as much as 20 kilos of stores. During this time, your colony is very vulnerable and open to attack by predators. Small mammals, such as shrews, can enter the hives, as can smaller birds. Woodpeckers can knock holes in the wooden hives and other animals can also harm the wooden frames. Every hole, every crack means a significant loss of heat, which the workers will not be able to compensate for in case of doubt. The colony will freeze to death.

Varroa mites also pose a threat during this period. They weaken the bees by sucking hemolymph. Weakened animals are less able to work, need more food and are more susceptible to diseases transmitted by the mite.

It's getting serious - work on your bee co-lony

When working around the bee colony, you as a beekeeper are usually standing directly in front of the open hive and have direct contact with the bees. It can happen that the bees see you as a danger and react aggressively. Therefore, in addition to suitable protective clothing, calm and planned work is particularly important.

Be clear about the tasks ahead of time to minimize disruption to the bees. Work slowly and calmly, work cleanly and always according to a strict plan. You should absolutely avoid working on several colonies at the same time, not least because of the strange smell, the bees could rightly react aggressively to you.

Work on the hive is always carried out only in fine, dry weather.

Before you start, make sure that you know exactly what work is coming and that you have prepared all the necessary utensils that you need for this work. Here it is also recommended to have created enough space. When you lift frames, do not place them on the floor, for the sake of good hygiene. Remove a frame, you should have the comb stand ready.

So you see: working without thinking will not get you anywhere at the hive and will only cause hectic and stress, which your bees will notice and react accordingly.

SPACE MANAGEMENT

To get through the winter well, bees prefer a shelter with a volume of about 40 liters. This small space they can heat up quite well and keep warm. If it goes over to spring, the colony starts to grow. Now you as a beekeeper must become active to prevent swarming. You give your bees more space. The advantage for you is at the same time that a large colony can bring in more honey. You overwinter your colony single-barrel, with only one frame, floor and roof. In the summer, a second frame is added. The best time to place the second frame is when the colony fills the entire first frame. If the colony also fills the second frame, it has reached the size of a commercial colony. Only now, at this size of the colony, you can harvest honey without robbing your bees too much of their livelihood.

As the second half of the year and late summer approach, the bees begin to prepare for the approaching winter and reduce the size of the colony. September is the right time to remove the second frame and overwinter your colony single-kept. However, this can cause space problems for large commercial colonies. It makes sense to decide whether to overwinter with one or two frames based on the size of your colony in September. One frame is perfectly adequate for about 5,000 bees. Two frames provide space and winter provisions for a colony with 15,000 bees.

The more barren the landscape and the colder the winter, the smaller you should winter your prey.

PROPAGATE THE BEE COLONY

Planned propagation of the bee colony through the formation of a scion is done in order to prevent unplanned swarming. There are several different ways to form offshoots from your own colony. This book will introduce you to three of them. Generally, if you form offshoots earlier in the year, they have longer to grow and prepare for winterization. However, if you form them too early, an unexpected cold snap, such as often occurs in May, can be fatal for the still weak colony. They will not find enough food during this cold period and will also not be able to keep the hive at a constant temperature above 30 degrees.

Hatchery
The brood hive is transferred to a single-barrel hive. This scion consists of:

• Two forage or honeycomb

• One pollen comb

• Different number of brood combs (in principle, the number corresponds to the current month plus 1 comb)

- Frames, possibly with middle walls to make it easier for the bees

- A queen cell ready for hatching

Brood combs are removed with the resident bees, i.e. the bees that are sitting on the combs at the time of removal move with them. It is possible to remove brood combs from different colonies, provided that they are all absolutely healthy. It is also possible to create a brood hive with only one brood comb. In this case, however, the small colony requires extra care, and it may be advisable to choose a smaller housing for the start to make it easier for the bees to heat.

Do not place the brood hive near the original colony, otherwise the bees will make the return journey there and will not stay in the new hive you have formed. Once the colony is established, you can return the hive, now the bees belong to the newly formed colony. You can recognize this very easily when the queen begins to lay her own eggs.

The queen is ideally placed in the brood hive with a queen cell that is ready to hatch. A hatching queen is always accepted, but if you put a young queen into the colony, she is often rejected. If it is possible for the bees, they will take their own queen from a re-creation cell. It is also possible to take the queen for your brood hive from the old colony. In this case, however, you must not mix bees from different colonies. The queenless colony will then obtain a new queen from replenishment cells.

Artificial swarm

Another way to form a scion is the artificial swarm. This way you simulate swarm formation and your colony is forced to make a completely new start. That is, you take only the bees into a fresh, empty housing. Honeycomb construction must now start from scratch; the first stores must be collected and finally the first eggs must be laid.

This offspring formation is also suitable for the rehabilitation of an unsanitary or badly aged bee housing. They do not take over any combs or stores. This can sometimes be the last chance to save your colony from the varroa mite. If there is an infestation, you can carry out the treatment against the varroa mite in the first days after the move.

To relocate a colony in this way - or to form a scion - sweep the bees from the old combs into the new housing. To ensure the supply of the bees in the first days, it may well be necessary to offer forage combs. For this purpose, feed dough, liquid feed or honeycombs from clean hives are suitable. In an artificial swarm, the queen is usually added by hand. To prevent the queen from being repelled, place the queen in a kind of cage in the new housing. This cage is closed with a plug of food, which is consumed after a few days and the queen is released. During these days, the colony and queen have time to get used to each other and start first contacts through the cage.

It may happen that the bees are not satisfied with the

new home you have given them and want to move out again directly. You must avoid this at all costs if you do not want to lose the swarm. To make sure that the bees get used to their new home, close the flight hole for one or two days. It is quite easy to explain why an artificial swarm needs a few days to get used to the new housing, but a naturally formed swarm does not have this problem. Before a colony swarms, it goes through several stages of preparation. Among other things, the wax glands of the workers are reactivated, the bees fill their honey bladders with provisions and the queen is prepared for the flight.

The artificial swarm does not have this time of preparation, so it needs a few days to activate its builder bees. Once the first combs are formed, the swarm settles down and the need to move out again is gone. You may remember the danger of the closed flight hole: the colony can easily overheat because no cooling can take place. So be sure to put the hive in a cool place during this time. From this has developed the term "cellar detention" as a synonym for this period.

Division into Flugling and Fegling

The division of your colony into flightling and fegling always happens on a day of peak activity in the hive. You take advantage of the fact that almost all collectors are on a foraging flight and you remove the old hive from its place.

In the exact same place, so that the collectors can also find their way back, place a new hive in which you hang some brood cells from the old colony, as well as two food or honeycombs. Be sure to use brood cells that contain mostly capped brood that will soon hatch. This part of the colony is called a flightling because it now temporarily consists almost entirely of flight bees. A queen must be added to the flightling. You do this either by adding a queen cell that is ready to hatch or by adding a queen in a cage.

The flightling is not able to raise a queen by replenishing cells. What is possible, however, is to add the old queen to the flightling. The bees that remain in the old hive form the fegling. The reason for this is that the bees that you take out with the brood combs for the flightling are swept back into the old colony. In the hive of the fegling you close the gap created by the removed brood combs by moving up the brood combs. Make sure that there is enough food in the colony, because the fegling contains almost no flying bees for a short time, which could bring in tracht. The fegling can take care of a queen itself by creating new cells. Of course, you can also leave the old queen in the colony or add a queen cell that is ready to hatch.

The creation of a queen is only possible if uncovered brood or eggs are present in the brood trays. Capped brood is already too old and can no longer be raised to queens by the workers. Therefore, it is not possible for the fly to form replenishment cells, because you yourself have taken care to insert capped brood combs, so that the colony quickly gets young offspring.

ADDING THE QUEEN

If you want to create offshoots from your colony, in certain cases it is necessary to add a queen, if your colony cannot grow its own new queen. This is a rather difficult process, as foreign queens are accepted only reluctantly. By providing suitable conditions, you can increase the probability that the new queen will be accepted.

Make sure that the colony, in which you want to put the new queen, has no open brood cells. Capped brood is allowed. Also make sure that there is no queen cell in the colony.

Young, mated queens have a better chance of being successfully accepted than if you add an unmatted queen to the scion.

Do not simply place the foreign queen in her new colony, but protect her in the first few days by using the cage already mentioned, also called the feeder cage. This gives the bees time to get used to the smell of the new queen and to make first contact with her. After you have hung the cage in the colony, you should leave it alone for the time being. Observe it during this time only from the outside, without opening the hive. Are the bees calm, is the work going well? After about two weeks you can dare to look into the hive. A sure sign that the colony has accepted the queen is the presence of fresh brood.

Breeding your own queens is only recommended for experienced beekeepers, as queen breeding is not easy.

However, you can ask your beekeeper association where you can buy healthy queens. These queens are usually color-coded, so you can make a note of exactly how old each queen is.

PLANNED REWILDING

The process of reversion refers to the replacement of the old queen bee with a new one. The bee colony does this automatically in certain situations. However, there are also situations in which you, as a beekeeper, want to replace a queen specifically. For example, you may want to do this before the old queen's egg-laying performance begins to decline, in order to keep your colony strong. Another reason to replace your queen is an aggressive colony. Increased aggressiveness can be genetic. By adding a queen bred for gentleness, you can influence behavior.

In order to transfer a colony according to plan, you must first of all remove the old queen from the colony. Colonies without a queen, without a queen bee cell and without the chance to form new cells will accept a new queen offered by you more easily.

Wait several days after removing the old queen and then take a look inside the hive. If it is your declared plan to integrate a purchased queen, you should remove existing queen cells or post-creation cells at this time. If you do not find any queen cells or replenishment cells after these days of waiting, it is likely that there is still a queen in the colony that you need to find first. Once all queens, queen cells and replenishment cells have been removed, you can hang your new queen between the combs in the add-on cage and start the integration process.

HANGING HONEYCOMB

According to the textbook, the brood nest is always located in the center and is surrounded on the left and right by several combs that serve as honey storage. Do not be alarmed if a different picture presents itself when you open your hive. Of course, you can now move the brood nest and place it in the middle of the frame, but you can also leave it where it is, according to the motto: "The bees will already know what they are doing". What you ultimately decide always depends on your colony and the circumstances.

However, if you decide to move the brood nest, you should always do so only as a whole and not swap the combs. Otherwise, you run the risk of tearing the brood nest and unnecessarily confusing the bees and disturbing their work flow. The process of rearranging combs is called correction of bee fit. It may happen in certain situations that empty combs hang between them and, for example - this often happens in spring after winter - an empty comb separates the brood nest from a honeycomb hanging behind it, which still contains plenty of stores. The bees have lost contact with the full comb and are starving, even though food is still available. In this case, it makes sense to correct the honeycomb. Restructuring can also make sense if you want to hang new combs, for example, to create more space for the brood nest or to hang a drone comb in the spring.

FEEDING YOUR BEES

Bees do not produce honey for humans. They produce and store only as much as they need as a colony to get through the winter, feed the brood and survive. Humans are not provided for in this and every honey you take from the hive, the bees pay dearly with too little food. Therefore, it is imperative that you feed your bees to ensure their survival. Never take all the honey out of the hive. Honey that is near the brood nest remains in any case as a fast-moving supply for the bees to bridge bad weather periods.

Winter feeding

For winter feeding, liquid sugar solution is used, which you offer to the bees inside the hive. A suitable time for feeding is late afternoon/evening. Now the bees retreat into the hive for the night and there is less danger of neighboring colonies coming to raid. You now put an empty frame on the hive and place the liquid feed in a bucket on the top frames. Place straw, cork or similar material in the solution so that the bees have a place to sit at all times, otherwise they will drown while feeding.

You prepare the sugar solution itself in a mixing ratio of 2 to 3. Two parts fresh water and three parts commercial household sugar. The sugar takes a while to dissolve in the water, because you must not heat the solution, which would accelerate the process, but would produce hydroxymethylfurfurate (HMF), which is harmful to the bees. Therefore, you should prepare the solution the night before and stir occasionally.

Emergency feeding

If there is a hive gap in the summer, a large colony may have severe problems surviving this gap. The previously built up high activity and growth rate of the colony requires a constantly stable food supply. If the hive is now missing, the stores are depleted and the bees begin to starve. To prevent starvation, you should now carry out emergency feeding.

For emergency feeding feed dough is used. You can either buy feed dough in stores or make it yourself. To do this, mix five kilos of powdered sugar with one kilo of your own honey. If the dough becomes too dry, add a dash of water. Note that no honey may be harvested after an emergency feeding this year if it takes place in the summer. Emergency feeding in the spring requires you to have high knowledge in order to carry out a honey harvest in the same year, because you are not allowed to harvest stored feed.

A good option is to hang honeycombs in the hive when there is a shortage of food. So if you still have honeycombs in reserve, you should always use them up first, because then you can harvest honey without any problems.

CONTROL YOUR HONEY BEES

As a beekeeper, you take responsibility for your bees. Now you have to check regularly if your animals are doing well or if there are problems. In winter, it is quite sufficient to carry out a check every 2 to 3 weeks. Here you reduce to the control of the mulch. If this is seasonal, then a close inspection of each individual comb can be omitted. This is especially practical in winter, as you run the risk of undercooling your bees by pulling out the combs.

In summer, inspections should be more regular and thorough, especially to detect and prevent swarming at an early stage.

You should carry out more thorough inspections, especially in the spring and into the fall. These inspections are called spring inspection and fall inspection.

You record every inspection of any kind in your stock cards. In this way, you will have a complete documentation and the possibility of later retracing your actions. The hive cards also serve to keep the bee colonies healthy.

Garbage
The bottom of magazine hives has the possibility of inser-

ting a bottom board made of wood or other material. This board is called a bottom board, a garbage diaper, a bottom insert or a diagnostic slider. The board is not left in the hive permanently because it attracts parasites and predators. Bees are also very clean animals that clean up their garbage to keep the hive clean. So, to prevent the bees from cleaning the bottom board, place a mesh between the hive and the bottom board that the bees cannot fly through. After about three days, remove the garbage board and the grid so that the bees can resume cleaning the floor.

Garbage is especially suitable for diagnosis and review in cold periods, when you do not want to open the hive. In spring, as soon as the first warm days arrive, you can insert the diagnostic slide. If you find colorless wax platelets on it, this is a sign that the builder bees are active and covering the brood cells. Your colony has started to brood. If you find small brown wax crumbs with fine fibers sticking to them, this is a sign of the first hatched workers. The fibers are the remains of the cocoon. The best way to identify and look through the junk is with a magnifying glass.

At the end of March, you can easily estimate the location of the brood nest based on the garbage pattern. It allows conclusions to be drawn about the size and activity of the brood nest. If you find white, almost translucent crumbs, you have a safe statement that winter food is still present. The crumbs come from opening the wax covers of the food combs.

The size of the brood nest and whether the bees are still actively engaged in brood production can also be seen in the autumn from the colony's droppings. At this time, and as winter progresses, the hive's brood can also tell you about the parasitic load on the colony and whether predators are tampering with your colony.

Parts of dead bees in the garbage provide evidence of intruders, such as the shrew - an insectivorous rodent - or wasps and hornets. Hornets use the protein-rich parts of the bee - the thorax with the flight muscles - to feed their brood. The status of the varroa mite in the hive can also be read off from the droppings.

The spring review

When it is finally warm and sunny, the time has come for spring inspection. In the process, pull out the combs one by one and inspect each of them very carefully. Always remember to do this only if your colony can not freeze to death in the process! The outside temperature should be just below 20 degrees, the day must be sunny and windless. Do not be hectic during the spring inspection, but still keep the work as short as possible. All the utensils you need - smoker, sweeping broom, water for cleaning, comb rack for hanging the combs - are absolutely ready!

Since you have every comb in your hand during the spring inspection anyway, take the feeder crown sample for American foulbrood diagnostics at this time as well.

Spring inspection coincides with the change of generations within the bee colony. The winter bees die, the first summer bees already hatch. Pay attention here to the quantity and quality of the brood cells. Are there enough pins, open and capped brood?

Inspect damage caused by intruders during the winter. Remove feeding traces on honeycombs or wax moth webs and larvae, moldy combs and the like. Fill the resulting gaps with empty combs or center walls. Also feed combs, which you inserted in the autumn, are removed now, even if still remains of unused fodder are contained in it. Otherwise, the feed stored in them can be transferred to the honey chamber and distort your honey harvest.

If your colony did not survive the winter, you should take a thorough look at the disaster. Hold a necropsy and try to find out the causes. Did the colony freeze to death, starve to death? Were there intruders? The cause usually lies in small or large mistakes during wintering, which must now be fathomed in order to avoid them next year.

Swarm controls

Swarm checks should be carried out weekly from the end of April to the beginning of July. With a healthy economic colony you can assume that it is strong enough to prepare for swarming.

Warning signs that your colony is thinking about swarming can be recognized by a thorough inspection of the combs. Experienced beekeepers use the so-called tipping method to check for swarming behavior, but as a beginner you cannot easily do this yet. The inspection is more precise and thorough.

What you can see on inspection when your colony is preparing to swarm are game cups and swarm cells (queen cells). Game cups are unpinned cells that the workers create but do not yet use. A queen cell then becomes a queen cell as soon as the queen has laid an egg in it.

If you find game cups and open queen cells during inspection, you should crush them every time. A closed queen cell is an alarm signal. Do not destroy them. After mating, it takes only six days for the new queen to hatch. If you find capped queen cells, the old queen has usually already left with a swarm. You have missed the swarm. This is not the end of the world, but you should definitely leave the queen cell in place at this moment, because your colony needs a new queen.

Signs that your colony still has a queen are freshly set pins in the brood nest. About three days before swarming, the old queen stops laying eggs. So if you don't find any of these fresh pens, it is likely that your queen has already moved out. Now all that's left for you to do is check the surrounding trees. You may still be able to find and capture your swarm.

Thorough autumn review

The last warm days of fall are your last chance to schedule a thorough walk-through to find and fix any final problems before winterization. Issues you should be sure to address with the fall walk-through are:

• The size of the volume: is the space enough? Do you need to reduce the size of the space, if necessary?

• The health of the people: do you find signs of disease?

• Food stocks: Are the stocks that the people have built up themselves sufficient, or do you have to feed in addition?

• Varroa mite: How severe is the infestation?

• Winter measures: Is the colony winterized?

The colony's space requirements must be so small in winter that the colony can keep the hive at operating temperature with little effort. If the bees have to make a lot of effort for this, the consumption of the stores increases too much. So reduce the brood chamber and estimate from the still active brood what strength your colony will have once the queen stops laying eggs for the winter.

The health of the bees can be determined on the one hand by the animals themselves and on the other hand by their nutritional condition. If you find dead bees in the colony or even pollution, this is always an alarm sign. As you have already learned, bees are very clean animals that do not tolerate pollution in the hive. In this case, investigate the cause, ask your beekeeper mentor and a veterinarian who specializes in bees for advice.

In August you carried out winter feeding to give your bees the chance to easily store stores. You now use the autumn inspection to check the actual stocks available. These will depend on the conditions. It may be that the bees have used up a lot due to gaps in the hive, rainy days or the like. On the other hand, a bee-friendly late summer and early fall may have brought in more tracht, so that large forage stocks are now available.

A commercial colony that you want to overwinter two-armed, requires 18 to 22 kilos of supplies. For a single-arm scion, 12 to 15 kilograms of stores are sufficient for the winter. A fully-filled storage comb according to Deutsch Normal Maß is equivalent to two kilograms. By Zander measure, it is equivalent to about 2.5 kilograms.

The flight operation of the bees is completely stopped in October to November, depending on the weather. By then you should have replenished the amount of feed. If it is necessary to offer feed again, you should do this before this time! More detailed information regarding the varroa mite can be found in a later chapter. Here it should be briefly said that the autumn inspection is considered the last possibility to assess the infestation with the mite. For this purpose, you can refer to the inspection of the brood. Varroa mites infest the brood and - if there is no brood - the adult workers. Treat with oxalic acid in the fall, when the colony no longer has brood.

Protective measures around the winter must now be checked and installed. The flight hole is the only way into the hive. In autumn and ultimately in winter it is ungu-arded as the bees have retreated to the winter cluster, and predators can easily enter and leave. Already at the time of winter feeding, you can reduce the size of the flight hole by using flight hole wedges and thus curb predation among the colonies.

Against shrews and smaller birds, you can now additionally attach a mouse grid or a net. Above the prey you can stretch nets to keep woodpeckers and ravens away from the people and especially their dwelling, so that no hole can be made in the outer shell. Secure your hive from fall and winter storms so that it cannot be knocked over and destroyed.

Winter controls

Checks during the cold season are carried out exclusively from the outside. Here you regularly check whether the protective measures are still intact. Has the mouse grid shifted and the net still properly lashed? The prey itself is also inspected to see whether storm or other impacts have caused the frames to shift and whether they are still properly seated straight on top of each other. Possible damage caused by wild animals must be repaired immediately. Depending on the region, raccoons may also attack the prey and in not uncommon cases the prey is knocked over and broken open in the process.

Broken open hives are usually no longer salvageable, since the colonies cool down abruptly and die quickly. You can try to save a freshly opened hive by immediately insulating it. Close all holes to avoid drafts and the penetration of cold air.

DOCUMENTATION OF YOUR WORK

Part of developing good beekeeping practice is learning from your mistakes and not repeating them. After all, you want only the best for your animals. The best way to recognize mistakes is to have functioning documentation. Especially if you take care of several colonies, the hive map, which is kept for each hive individually, makes it easier for you to read and remember certain conditions, events and work carried out.

Stock card

Everything that happens around your colony is entered here. Data and facts about the queen, the mite treatments, the results of the mulch and the spring and fall inspection, etc.. Everything that seems important to you and everything that seems unimportant is noted in the hive card.

Honey book

In the honey book you record everything about your own honey, so that for each jar it is traceable when it was harvested, when it was centrifuged, bottled and stored. The batch numbers, which you must assign as a honey seller, are also noted in the honey book.

Inventory book

The use of pharmacy-only medications must be documented according to the Animal Keeper Medication Verification Ordinance. These pharmacy-required medications also include the acids used to combat the Varroa mite. In order to keep track of all veterinary and curative treatments that you administer to your bees, it is worthwhile to also record non-pharmacy treatments and applications in the stock record book.

The honeycomb and the wax

In the wild, bees regularly change their housing, leave the old combs behind and start building new ones from scratch. Among other things, this has hygienic aspects, because pathogens and parasites accumulate in the wax of the used combs.

Wax, which comes fresh from the wax glands of the worker bees, is colorless and whitish translucent. It acquires its typical yellowish color through contact with pollen, nectar and the bees themselves. The pollen in particular is a true coloring agent and quickly causes the wax to lose its pure white color. This is not yet a problem, since it is only the storage of fat-soluble dyes.

Contamination occurs in the brood combs due to the larvae's legacies and, of course, the pupa remains. The cleaning bees try to clean and disinfect them with their saliva and the use of propolis. Propolis has a dark brown coloration, which is gradually transferred to the brood cells, which are then deep dark brown in color. The cleaning bees do not succeed in cleaning the honeycomb 100% again, remains of each brood remain in the honeycomb. This organic material is a good breeding ground for bacteria, germs and parasites. The wax moth, for example, is one of these parasites. The lesser wax moth lives in the brood cells and bores tunnels through the wax from brood cell to brood cell, where it feeds on the leftovers and debris.

Bacteria found in the organic debris are sometimes infectious for several years and lead to constant reinfection of newly hatched bees. This is also one of the reasons why you should never swap combs among colonies.

DETECT AND REPLACE OLD COMBS

Your job as a conscientious beekeeper is to replace and renew the combs regularly. To do this without losing brood, you can take advantage of your bees' behavior.

Long-term honey stocks are always stored away from the fly hole, i.e. in an area that is located in the upper frame of two-barrel colonies. The brood nest is located in the middle. In autumn, when the colony reduces in size, it usually moves to the upper frame, so that the combs located in the lower frame slowly empty and can be taken out by you.

Honeycombs from the last honey harvest are called honey-moist empty combs after spinning. You can reuse these as new combs for the brood nest. On the other hand, former brood combs are in no case suitable for honey extraction and should be disposed of.

To avoid mixing of the honey room and brood room, you can use a grid through which the queen will not pass.

This way she cannot lay eggs in the honey room and the combs there are used only for storing ripe honey.

If your colony reduces in size, remove this grid and the brood nest can move to the former honey chamber, the previous brood combs become free and are ready for disposal. By rotating the brood chamber in this way, the bees give you a sensible cycle for honeycomb renewal. Ultimately, it is more hygienic to replace combs sooner rather than later to contain disease and parasite infestation.

WAX

Sort out about 10 combs per year in each of your colonies. You do not reuse dark wax from brood combs to cast center walls. Thus, you would only hang the bacteria contained in it back into the hive in another form.

There are companies and beekeeping supply stores that accept old wax and exchange it for middle walls. If you would like to accept such an offer, inquire about the exact conditions beforehand. In most cases, these companies accept the old wax only after it has been removed from the frames, which means that you must also remove it from the wires. It is only worthwhile to exchange your used wax for middle walls if they are manufactured without residues, otherwise you run the risk of buying diseases. A corresponding certificate proves the absence of residues beyond doubt.

Melting wax yourself

If you enjoy making your wax into candles, for example, it's worth buying a sun wax melter. These come in various sizes with space for 2 to 3 combs.

In your sun wax melter you can melt wax on the one hand, which you have cut out of fresh natural construction. The wax that your bees create in the natural nest is still very clean at the beginning. If you catch it before there are eggs and larvae in it, you can melt it down and use it for casting or pressing midwalls. The equipment you need to do this is usually not worth it for a beginner. Perhaps your beekeeper sponsor has the equipment or you can ask at the beekeeper's club.

The wax from the old brood combs is not suitable for reuse in the colony, as mentioned, but you can use it for candle making.

On sunny days, melting wax from the natural structure takes about 1 to 2 hours. Brood combs need a little more time. Fill the collection vessel on the melter with a little water beforehand, so you can collect the wax better. After the old combs are melted, you may find residues of cocoons and larval droppings. Larvae of the wax moth may also appear.

Wax clarify

Collect the raw wax obtained from the melter separately old and fresh wax. Once you have a certain amount together, you can heat the wax with the same amount of water in a large coated pot. The pot should be coated, as wax turns gray when it comes in contact with iron. Also, use an old pot - you can't use it for cooking afterwards. Heat your wax-water mixture to about 85 degrees so the mixture doesn't start to boil and produce hot wax splatters. Stirring it slowly several times will loosen any residual contamination from the wax. The very slow cooling that now follows causes a dirt zone to form at the boundary between the water and the wax, which you can remove later with the stick chisel.

Making candle wax

You can now use this simply clarified wax for further processing into candle wax. Heat it again until it is liquid. To prevent your wax from burning, stay with it throughout the process and it is best to heat the wax in a water bath rather than directly in a pot. In the next step, strain the wax through a fine-fibered filter that will catch the suspended particles and minute remnants of feces, cocoon, and chitin shells.

The wax is now still quite a dark color, which you can lighten by various means. You can pour thin wax plates and hang them in the sun for several days. This will give the plates a much lighter color. Another way is to lighten them using acids or alkalis. As a beekeeper, you already have formic acid or oxalic acid at home, since you need these agents to fight the varroa mite. But you can also work with citric acid, which is used in most households as a natural decalcifier.

Always wear gloves, safety goggles and appropriate

protective clothing when working with acids. Otherwise, splashes may occur during heating, which can seriously injure your eyes.

To lighten the wax, put it in a saucepan again together with water. For about five kilos of wax, you need one liter of water and, in the case of citric acid, two grams of citric acid. Heat the mixture to 85 degrees and stir constantly. A mechanical stirrer is also suitable, since water and wax must come into close contact. Sheathe metal stirrers or use wooden stirring sticks.

To test whether your wax has already taken on the desired color, dip a wooden stick into it and let a thin layer of wax harden on it. In this way, you can optimally judge the color. When cooling the finished mixture, again make sure that it cools very slowly so that the water and wax can separate properly. If in doubt, you can wrap a blanket around the pot as insulation to slow the cooling down again.

The basics of hygiene

Honey is a natural food that is consumed raw. Bacteria, impurities or pollutants are thereby eaten by people. This presents the beekeeper with the challenge of working absolutely hygienically from the beginning to the end of the honey production chain.

If you want to sell your honey, you must inform yourself in advance about the legal regulations on food hygiene. Ask your beekeeper sponsor and your beekeeper association about this, the legal regulations are very fast-moving and change from time to time, so it is not useful to write them down here.

Most contamination is not caused by the bees, but by the beekeeper himself, who introduces dirt and germs through unclean work on the hive or when spinning and processing honey. The centrifuging room itself must also meet certain hygiene standards. Last but not least, waiting periods and instructions in connection with varroa mite control must be observed.

Make sure that your work clothes are always clean and that you wear different work clothes at the colony than when processing honey. Think of clean, food-safe disposable gloves and a hair net! After each exit of the room the protective clothing is renewed absolutely.

Remember never to put combs and frames on the ground while working on the colony and always have clean water ready to thoroughly clean your tools and hands in between.

CLEANING AND STORING YOUR EQUIPMENT

Honey is a particularly sticky and tenacious substance. Make sure to clean all your utensils with water without leaving any residue. Half-removed honey residues are true germ havens! The dishwasher does an excellent job here. Run it at 60 to 65 degrees to clean and disinfect honey jars and lids.

You also need to keep clean hive and frames. After removing all wax residues from the frame, remove the wires and dispose of them as well. Place the frame in 2% sodium hydroxide solution for comprehensive disinfection.

THE OWN WORKSPACE

If you extract honey only for your own consumption, so you can use your own kitchen for further processing. As a hobby beekeeper, this is an obvious choice. If you want to have your own room for your honey extraction, it must meet certain requirements, which are clearly defined in the relevant legal texts. These include the following conditions:

• Neat, tidy condition

• Dry, no formation of accumulated moisture

• Intactness: cobwebs, cracks, crumbling plaster, damp spots are not allowed

• Substances hazardous to health must be stored outside, e.g. cleaning agents

• Free from plants and animals, fly screens on open windows.

When setting up your own room, pay attention to the regulations that apply in your region. In order not to do anything wrong here or forget essential things, you should ask your beekeeper mentor for advice in case of doubt or get tips from your beekeeper association.

Medical considerati-on

As with humans or other animals, preventing the development of a disease is better than treating it. So make sure you keep your bees as healthy and strong as possible to minimize the need for treatments. You keep your colony strong and healthy by providing them with optimal conditions, through a well-chosen location, for example. Check your colonies regularly for disease. The sooner you detect a deviation, the more promising medical treatment will be.

Do not underestimate how easily and quickly you as a beekeeper can transmit a disease from one colony to another. Therefore, pay special attention to hygiene. This includes a clean beekeeper's suit, clean equipment, fresh water and hygienic work itself by not allowing frames or frames to come into direct contact with the ground. Replace old combs early to maintain a high standard of hygiene within the hive as well.

To limit the spread of diseases as much as possible, you should not exchange frames, frames or combs between individual colonies. It is also not advisable to combine colonies without first ensuring beyond doubt that both colonies are absolutely healthy. Freshly purchased colonies should also have an undoubted health status and additionally spend a certain time in quarantine. Also make sure that each individual colony has enough space. The more densely you place your colonies, the easier it is for diseases to be transferred from one hive to the other.

VARROA MITE - VARROA DE-STRUCTOR

The most widespread parasite of bees is the Varroa mite (Varroa destructor), which has spread rapidly in the last twenty years. Nowadays, there is hardly a colony that is not infested by the varroa mite. The aim of the treatment is therefore to contain the infestation to such an extent that it does not affect the bees and does not pose a threat to the survival of the bee colony.

Three-part integrated treatment concept
The integrated treatment concept was developed and tested by bee institutes and it provides a well effective therapy. It was developed for the treatment of commercial colonies, so care was taken not to affect the quality of the honey. This must be absolutely residue-free. Only organic acids are used, which do not harm the bee or the honey, do not accumulate in the organism, but are completely degraded. This is achieved by using substances that occur naturally in the bee. The use of the acids after the honey harvest guarantees the freedom of the honey.

Since no honey is harvested from the hives yet, you can determine the time of treatment here exclusively according to the infestation situation. Formic acid or lactic acid are usually used here. If in doubt, ask your beekeeper mentor for advice, as the treatment of the Varroa mite should always be individually adapted to the respective colony.

To know if your treatment has worked, always perform a success check. The easiest way to do this is to look through the garbage. For this you determine the natural dead fall of the mites per day before the treatment. Then you treat and determine during the treatment the dead fall of the mites by the active substance. Some time after the treatment you determine again the natural dead fall of the mites per day. Ideally, at this point you will not find a single mite left. Your colony is - until reinfection - mite-free.

The first step of the concept is cutting the drone brood in spring. Part two consists of treating the brood with formic acid in midsummer and fall. The third part is the treatment of the workers in winter with oxalic acid.

1. cutting the drone brood

Drones need 13 days for their metamorphosis, which is on average two days longer than a worker bee. The varroa mite knows this difference and prefers to attack drone combs, as more mites can develop there. You can take advantage of this fact. Hang drone frames in your hive. A drone frame is an unwired frame in which the bees create drone cells in the wild. The center walls that you normally provide in the frames are used by the workers to create smaller honeycombs for workers. Drone combs are only possible in frames without a center wall.

For small colonies, hang the drone frames on one side next to the brood nest; one frame is sufficient. For commercial colonies, hang one drone frame each to the left and right of the brood nest between the honey stores and the brood nest.

The drone frames usually have a wooden bar in the middle that divides the frame horizontally into an upper and lower section. This allows you to cut out the brood in portions. You now always cut out the already capped brood before it hatches from the comb. Uncovered brood remains in the hive so that the female mites can lay their eggs there. Before the mites hatch at the same time as the finished drones from the capped combs, they are now removed and frozen for 24 hours to safely kill the mites. Smaller quantities of this waste can be disposed of in household waste.

2. treatment of the brood with formic acid

Bees and mites have a different metabolism, which is why the application of a 60% formic acid solution is completely harmless for the bees, but has a killing effect on the varroa mites.

Treated about the wet heather evaporator. Here there are several models, of which the model "Professional" is the most beginner-friendly and reliable model. The application via the evaporator guarantees you the correct concentration of formic acid, so that you do not accidentally harm your bees.

The appropriate time for application is immediately after the second honey harvest, i.e. the last honey harvest of the year, in a commercial colony. The application is then repeated again in September. This second application should be as close as possible to the winter dormancy, but should always be done when the outside temperature is still above 12 degrees Celsius.

The Nassenheider evaporator can be hung directly in the brood nest in an empty frame. Usually, however, an empty frame is placed on top and the evaporator is placed on top of the frame. Here, make sure to separate the empty space with a grid or net so that the bees are not tempted to go wild there.

Fill the evaporator with 200 milliliters of the 60% formic acid and insert a wick. Now place a piece of foil and a fleece on the upper frame. Place the evaporator on this fleece. The wick must have contact with the fleece, so that the fleece is slowly moistened with formic acid. Approximately 20ml of acid is evaporated daily. The duration of the application is at least 10 days in each case. Check the evaporator after 5 to 6 days and top up with formic acid if necessary. If more than 30 ml is evaporated daily, replace the wick with a smaller one to reduce the amount.

3. treatment of the workers with oxalic acid.

In winter, evaporation of formic acid is not possible, but neither is taking out the combs to spray lactic acid due to the outside temperatures. However, in a mild winter, the time between the last formic acid treatment and spring provides perfect conditions for the mites, as the bees continue to have brood in the colony, so it is necessary to reduce the infection pressure once again in winter and treat your colony with oxalic acid.

This can be done only at the time when the colony has no more brood. For the treatment, mix a sugary oxalic acid solution shortly beforehand. To do this, use 3.5% oxalic acid dihydrate and add sugar just before the application. It is very important to add the sugar just before, because HMF (hydroxymethylfurfural) is formed in a ready solution when stored for a long time, which is harmful to the health of the bees.

Now drip the finished solution onto the bees that are attached to the combs in the upper area of the comb alley. This way you do not have to pull out the combs and your bees do not run the risk of hypothermia. Through their cleaning behavior, the bees now clean each other of the sticky liquid, whereby the oxalic acid is distributed throughout the colony and can act against the Varroa mite.

Treat offshoots with lactic acid
You can treat smaller colonies against the Varroa mite with lactic acid. Lactic acid can only be fully effective in colonies without active brood, because acidification with lactic acid only works on adult bees, not on larvae.

A 15% lactic acid solution in water is applied. You

now pull each comb out of the hive individually, spray from both sides until all resident bees are completely wetted, and put back into the hive. Repeat the treatment several times after a few days.

Since farm colonies are used for honey extraction, you may treat them with lactic acid only after the last honey collection. Since active brood is still present in the colony, this reduces the effectiveness of the treatment. In winter, when there is no brood, the outside temperature is decidedly too cold. Since there is no honey extraction in the case of hives, you can repeat the treatment several times to reduce the mite pressure. It is not possible to be mite-free, as the colony also has active brood here.

You can treat swarms or artificial swarms with lactic acid before they start breeding. Here there is no active brood yet, so the treatment is sufficiently effective. However, you must choose the right time. If you open the hive before the swarm has formed combs, it will move out again. If you treat too late, the combs are already pinned and the first larvae have hatched.

AMERICAN FOULBROOD

AFB is a notifiable bee disease. This means that if you detect an infestation in your own colony, you must report this to the responsible veterinary office.

AFB is triggered by the spores of the bacterium Paenibacillus larvae. The spores infect the larvae of the bees and multiply in their digestive tract so that the larva eventually dies. Adult bees are not susceptible to the disease. However, they should not underestimate AFB, as a colony may well die if too many larvae fall victim to AFB.

For the control and detection of American foulbrood, contact your beekeeper mentor or a veterinarian specializing in bees. Your contacts will discuss the exact procedure with you and create a monitoring plan individually tailored to your colonies.

The honey

Honey - that's what you might be doing all this for. Acquire bees, beekeeping equipment, learn the craft of beekeeping - all for the honey! In order not to go beyond the scope of this book, honey extraction will be discussed only to the extent that you can expect honey as a beginner and self-consumer. If you want to sell honey commercially, you should contact your beekeeper mentor and the beekeepers' association to take an appropriate certificate of competence, as you then fall within the scope of the Food Code.

HONEY - WHAT EXACTLY IS IT?

The Honey Ordinance defines honey as follows:

"Honey is the naturally sweet substance produced by honeybees by ingesting nectar from plants or secretions of living plant parts or excretions of plant-sucking insects found on the living plant parts, transforming them by combining them with their own specific substances, storing them, dehydrating them, and storing and ripening them in the honeycombs of the beehive." Quote taken from the Honey Ordinance (HonigV), Appendix 1, Section I.

Honey is therefore always collected by the honey bee.

Already in the honey bladder, the addition of enzymes begins the ripening process of honey. In the hive, unripe honey is stored and transferred more often, each time removing water from it. This thickens and preserves the honey until it is finally ready to be harvested by the beekeeper.

Ingredients of honey

80 to 85 % of honey consists of sugar. Almost all sugar molecules are present as simple sugars (monosaccharides). The predominant remainder is water. Other ingredients make up only 2 to 3 %.

The sugar content is also called sugar spectrum and varies depending on the grapevine brought in. Three different sugars come from the nectar: cane sugar (sucrose), fruit sugar (fructose) and grape sugar (glucose). Absorbed cane sugar is broken down into fructose and glucose by the enzymes added in the honey bladder right at the beginning. The ratio of fructose and glucose to each other characterizes the honey and depending on how the ratio turns out, specific flower honeys can be determined. The consistency of the honey also depends on this ratio.

Honeydew honey can contain up to twenty different types of sugar. In addition to fructose and glucose, it may contain various two- and three-sugars, such as maltose, raffinose, isomaltose, meleziose, erlose and turanose.

According to the Honey Ordinance, the water content in honey may only be a maximum of 20%, and according to the German Beekeepers' Association, only a maximum of 18%. The low water content ensures the stability of the honey. Too much water causes the honey to spoil quickly.

The remaining 2 to 3% of the ingredients are various amino acids, proteins and minerals, as well as secondary plant compounds. In unfiltered honeys may be added traces of pollen and wax. Although the percentage is so low, these ingredients can affect the color and flavor of the honey.

Honey types

The Honey Ordinance prescribes various terms with which honeys are to be labeled depending on how they are obtained. Do not confuse the term with the designations honey variety/variety honey.

• **Spun honey**: This honey is obtained by spinning the combs. Nowadays it is the most common form of honey extraction.

• **Pressed honey**: The pieces of comb are wrapped in fine-meshed cloth and placed in a press. The honey drips out of the bottom of the press.

• **Honeycomb**: This honey is cut out of the honeycomb created in the natural construction. It must be free of brood. The piece of honeycomb comes with it into the jar and is sold with the liquid honey.

• **Drip honey**: the combs are uncapped and placed so that the honey can drip out.

• **Baking honey / cooking honey**: This honey is rather inferior in quality and must not be consumed raw. It is suitable as food only after heating.

Honey varieties and varietal honey - what is the difference here?

If you specify a type of honey, this is done without specifying a single grape source. For example, you can choose "spring honey" or "summer blossom honey" as a designation. Also "honeydew honey" is a honey type designation. Of course, the jar must also contain what is written on the front. This is for consumer protection.

You may call a honey a varietal honey, which consists predominantly of a Trachtquelle. More than 60% of the honey must come from the mass harvest, so that you may call your honey, for example, "rapeseed honey". Before you call your honey varietal honey, you must have a laboratory analysis carried out, because just because there is a rapeseed field blooming next door, it does not mean that your bees also use this rapeseed field as a source of honey.

Honey source - nectar

Nectar is produced by the plant to attract insects. The nectar is used to feed the insect, which also picks up pollen when it visits the plant and carries it from flower to flower, thus taking over the reproduction of the plants.

Depending on the type of plant, nectar production varies throughout the day. There is also a maximum amount of nectar per day, so that a flower can be drunk downright empty, if many bees were there on a visit. On warm days the nectar is somewhat more concentrated, if there is a lot of water available its consistency is more liquid. However, the amount of sugar is always the same in relation to the total amount of nectar per day.

Some plants have extrafloral nectaries, which are nectar collection sites outside the flower. However, these are almost never approached by the bees.

Honey source - honeydew

Honeydew is secreted by insects that suck plant sap. In Germany, these are mainly aphids and scale insects. Honeydew sources are mainly trees and shrubs, less herbs and infested perennials.

The aphids suck the plant sap (called phloem) to absorb the amino acids they need to live. However, since these are present in very low concentration in the phloem, the aphids have to suck a lot of it. To avoid bursting with water, they continuously release small droplets of honeydew, which contains not only water but also a lot of sugar.

Collect pollen

Pollen serves bees as a source of protein, of which a lot is needed, especially in the spring. Bees have a brush and comb on the inside of the rear pair of legs. They use these utensils to wipe themselves from front to back after a visit to a flower and use them to comb up the pollen that has stuck to their bristle coat. Once the fur is free of pollen, it is stowed in the pollen baskets on the outside of the hind legs. This process can only be done in flight, as the bee needs all its legs to do this, and the stripping into the basket is always done crosswise. The material collected with the left is therefore stowed in the right basket.

The process of collecting and stowing pollen is known as hilling in beekeeping parlance.

TRACHTPFLANZEN

Plants suitable for honeybees to collect must not have deep calyxes, as the bee's proboscis is not as long as that of the bumblebee or butterfly. Therefore, bees fly only to flowers with easily accessible flower bottoms or correspondingly shorter calyxes.

By now you are familiar with the term mass harvest. The opposite of this is the so-called Läppertracht, in which there is no plant species that dominates. Instead, the Läppertracht consists of many individual, different flowers that ensure the supply of bees even between the mass harvests and is therefore of great importance for your colonies. Below, we will introduce you to some native plants that have special significance for bees. If you have the chance to join a honey tasting, you should definitely do so. It is surprising how well you can tell the source of the honey from the taste of varietal honey.

• **Hazelnut**: Corylus avellana; one of the most important

sources of pollen in spring. Serves to build up the crop.

• **Willow**: Salix; Serves as a building crop, provides pollen and nectar. Pure willow honey is not available, as this crop is quickly eaten by the people.

• **Fruit trees**: apple, cherry, plum; possible in fruit-growing areas as a mass harvest, which provides a varietal honey, referred to as "fruit blossom honey". Considered varietal honey, even if several types of flowers have been entered. Aroma: discreetly floral; color: white to pale yellowish; consistency: creamy.

• **Canola**: Brassica napus; provides a good bulk crop in growing areas. Good pollen source. Varietal honey from rapeseed flowers belongs to the flower honeys. Color: light to pure white; Aroma: mild, sweetish; Consistency: firm-creamy.

• **Dandelion**: Taraxacum officinale; offers a large pollen supply and abundant nectar for build-up harvest. Varietal dandelion honey is possible depending on the region. Aroma: strong-intense, partly sharp-penetrant; color: golden yellow; consistency: initially viscous-thick, may crystallize.

• **Robinia**: Robinia pseudoacacia; bridges a possible gap in grazing between rapeseed and lime blossom. In urban areas, mass pollination and varietal honey are possible, often referred to as acacia honey. Color: watery-yellowish, sometimes shimmering greenish; aroma: fine, sweet, very sweet; consistency: very liquid (fructose out-weighs glucose, which is why the honey remains liquid).

• **Linden**: Tilia; excellent suppliers of pollen and nectar, especially in urban areas. Provide abundant honeydew due to the linden leaf aphid. Variety honey "Linden honey" as a mixture of nectar and honeydew honey. Variety honey "linden blossom honey" as pure nectar honey. Color: whitish, green-white to yellowish in lime blossom honey. Linden honey color: yellowish to dark brown. Aromatic strong smell of mint (menthol).

• **Clover**: Trifolium; possible as a mass crop in areas where clover is grown as a forage crop. Provides a-bundant nectar and pollen. Red clover in meadows is not approached by bees because their proboscises are too short. Other clover species such as white clover, incarnate clover, sweet clover and horned clover are used as forage. Variety honey "clover honey" possible. Color: light to white-yellowish; Aroma: floral, mild; Consistency: firm-creamy.

• **Heather**: Calluna vulgaris; the broom heather forms a mass crop in heath areas in August and September. Varie-tal honey "heather honey". Color: amber; aroma: spicy, tart, slightly acidic; consistency: gelatinous with fine and coarser crystals.

HONEY - WHAT THE BEE DOES

The bee collects honey for two reasons: immediate consumption or storage. Collectors bring the honey to the hive and hand it over here to the honey makers, who decide what to do with it according to their needs. You already know that directly in the honey bladder of the collectors by the addition of enzymes the honey preparation begins. If the bee arrives now in the hive, it can pass on the honey by Trophallaxis (the social feeding). This happens especially when little food reserves are stored in the colony.

If more food reserves are available, the bees empty their honey bladders in the feeding ring around the brood nest. The honey stored there is meant to be consumed soon, as it has a very limited shelf life as immature honey.

Honey that is not consumed in the next few hours or the following night is transferred by the honey makers. They suck the immature honey into their honey bladder, add digestive enzymes again and carry it into the honey room. In the process, they extract water from the honey again, so that the ripe honey has only 14 to 18% water content. It may be necessary to transfer the honey several times to achieve this water content.

Water extraction is possible very effectively in the honey bladder, through so-called aquaporins, which allow the water to flow from the honey bladder directly into the hemolymph. In the first phase, when the collectors transport the honey into the hive, the added sugar-splitting enzymes act. These require water for their work. Only in the second phase, when the honey makers then carry the honey around more often, is water no longer needed for the enzyme activity, and the sugar is completely split. Now the remaining water can be effectively extracted. This process turns about 2.7 liters of nectar into one kilo of ripe honey after water removal.

Once the process of honey ripening is completed, which takes several days, the honey can be stored in the combs for several months. For this purpose, the comb is covered with wax and only opened again when the colony has a need for food. Depending on the composition of the honey, it remains liquid for months. If it crystallizes, it must be liquefied again by the bees before use. Water is needed for this.

HONEY - WHAT THE BEEKEEPER DOES

As a beekeeper, you do not have to do anything more for honey ripening. Honey is ready as soon as you remove the capped combs. You harvest, extract and pack away.

Honeycomb removal

You should only use your smoker to a small extent to remove the combs, otherwise the honey may well taste like it has been smoked. Direct some smoke through the flight hole into the hive so that the bees calm down, and then start removing the finished combs. Gently sweep the sitting bees back into the hive or into a bucket if you plan to remove multiple combs, and hang the bee-free comb in a waiting transport container. A comb hive is only suitable for transport to a limited extent, as you need to ensure that your combs remain free of bees and that you do not carry any of the small insects into the house with you. A container with a lid, for example made of food-safe plastic, is more suitable here.

Immediately after extraction, the combs begin to cool down. Therefore, it makes sense to process the combs directly, because honey is more difficult to extract from cooled combs.

Discover

The next step is performed in the spin room or in your kitchen. To make your work easier, heat the room to at least 25 degrees so that the honeycombs do not cool down too quickly.

Make sure the hygiene of the equipment and yourself is impeccable! Now you have to open the honeycombs with the uncapping fork and the uncapping harness. Move the uncapping fork as flat as possible over the honeycombs and separate the wax covers. Slide these wax remnants into a tub to be melted down and reused later. Be sure to watch your fingers when uncapping. The uncapping fork has many sharp prongs, which can result in nasty wounds.

Therefore, always work in the direction away from your hand. As an alternative to the uncapping fork, you can work with a hot air dryer. This melts the delicate lids within seconds. Work on the combs only as long as necessary so that the heat does not penetrate deeper and damage the honey. This method only works if you have honeycombs in front of you that have never been used as brood combs - no problem, as this is essential for the hygienic extraction of honey anyway.

Spin

There are different types of extractors, which are either electric or hand-operated. For a beginning hobby beekeeper, a hand-operated extractor with space for three to four combs is perfectly adequate.

Centrifuges work by centrifugal forces and spin the honey out of the opened combs. To ensure smooth spinning performance, you need to make sure that the spinner is balanced, that is, balanced with combs. Thus, in a spinner with room for three combs, there must always be three combs. In a spinner for four combs you always work with four or with two combs, which in this case are opposite each other.

In the case of tangentially operating spinners, only the side of the honeycomb facing outwards is emptied. So in this case, start with a slow spin to empty the outside facing combs to a rest. Then turn the combs over, start again with a slow number of revolutions and increase the speed continuously until the outer combs are spun dry. Now turn the combs a second time and spin the first side completely dry.

To prevent honeycomb breakage, follow the instructions described above exactly. If you were to spin the first side dry at high speed, high forces would act on the still full side of the comb, which would quickly lead to comb breakage. Wires that are not tightly tensioned in the frame also lead to instability of the honeycomb and honeycomb breakage occurs more quickly.

Sieving and straining

Immediately after spinning, run the liquid honey through two sieves to remove wax particles, propolis particles, pollen and other impurities from the honey.

Start with the coarse sieve and filter the honey immediately afterwards through the fine sieve into the collecting container below the centrifuge. The fine sieve has a mesh size of 0.2 mm.

> **Tip**: During spinning, the coarse sieve may become clogged. Especially if larger wax particles are added, for example in the case of a broken comb. In this case, you must stop the spinning process and clean the sieve first to avoid overflowing.

Next, run the honey through a nylon mesh cheesecloth. This cloth has a mesh area of about 2 mm² and thus retains even the finest impurities. To keep the process of straining and straining effective, ensure that the honey temperature remains at 25 to 30 degrees, as honey that is too cold will quickly crystallize and clog the strainers. In this case, temper the honey, but be careful never to get to a temperature above 40 degrees or heat damage to the honey will occur.

Skimming

The finished sifted and strained honey is stored in a large container and stirred thoroughly once to combine the honey from all combs. Then close this jar airtight and leave the honey at room temperature for a few days. During this time, clarification takes place. The finest wax particles and air bubbles rise to the top during clarification and collect as the upper layer. Since the previously cloudy honey becomes clearer during this process, it is called clarification. Once clarification is complete, you can use a dough scraper or ladle to remove the top layer.

Crystallization

A saturated sugar solution will crystallize sooner or later. This means that the individual sugar molecules accumulate to form sugar crystals, which collect at the bottom of the solution. Whether and when a honey crystallizes depends on the ratio of glucose to fructose. Glucose crystallizes earlier than fructose. So, if your honey contains more glucose, crystallization occurs earlier, the best example being rapeseed honey, which hardens very quickly. If your honey contains more fructose, it remains liquid longer, as is the case with robinia honey, linden honey and honeydew honey.

Stir

Now, to get a solid honey creamy and effectively affect its consistency, you need to stir it. Every honey is different and stirring is an art in itself. It takes practice and some experience before you can get good at stirring and thus achieve the consistency you want. So don't be disappointed if it doesn't work satisfactorily the first time.

Liquid honeys, such as robinia honeys and honeydew honeys, do not need to be stirred, they remain reliably liquid for a long time. Blossom honeys from the early harvest should always be stirred, otherwise they sometimes become so solid that you would have to break them out of the jar.

When stirring, you create movement within the honey and friction between the individual glucose crystals. Glucose is present in unstirred honey assembled into large crystals that quickly precipitate (crystallize) and form hard compounds. These large glucose crystals are broken up and reduced in size by the process of stirring, so that the honey acquires a creamy consistency and remains "liquid" or creamy for a long time. To achieve this, the honey must be stirred during the time of crystallization. Once honey has hardened, it is difficult to make it creamy again. The right time to start stirring you have to read on your honey. To do this, fill a jar of fresh honey immediately after skimming and place it so that you can look at the jar every day. If the honey begins to cloud, this is a sure sign that crystallization is beginning. Now start stirring (or tamping) the honey. Each day, stir your honey for about 5 to 10 minutes, five days in a row. This should be enough to create a creamy honey.

Stir so slowly and evenly that the honey does not foam, but at the same time so quickly and thoroughly that there are no dead corners where the honey does not move. The entire honey must be moved. The stirring itself always takes place under the surface of the honey. Between stirring episodes, store your honey at a temperature of 10 to 20 degrees.

Summer honeysuckle inoculation

In the case of summer harvest honeys, it sometimes happens that they crystallize only very late. At this point, the honeys are usually already filled and it is no longer possible to influence the process. Therefore, you can accelerate the crystallization process by inoculating the honey with a trigger.

To do this, add 3 to a maximum of 5% foreign creamy honey to the finished skimmed honey and start stirring as soon as crystallization begins. Be careful not to add too much foreign honey, otherwise you will distort the character of the honey. This is not a big deal, but you will lose the characteristic flavor of the different honeys, which is also part of the appeal of making your own honey.

If you prefer a liquid summer honey, simply omit this step.

Filling

The jar in which you have stirred your honey ideally has an outlet tap at the bottom through which you can pour the honey directly into the jars. Make sure that the jars are undoubtedly clean and that you have not polished them with a tea towel or the like. This will get lint into the jar, which will then end up in the honey. Instead, take the dried jars straight out of the dishwasher. Label each of your honey jars, even if you are only making the honey for your own consumption. In this way, you will always know exactly when you have bottled, which trachtquelle is (probably) implemented in the honey and, for example, from which bee colony your honey comes.

If you produce for sale, you must comply with the legal requirements for labeling. You should obtain more detailed information on this by acquiring a certificate of professional competence.

Beekeeping calendar: your comprehensive checklist for the bee year

In the past chapters, you have now already learned quite a bit about the industrious honey producers with whom you would like to work. You were allowed to learn how the bees develop over the year, what processes they go through as a colony and what tasks are involved for you as a beekeeper (beginner). To make this a little more manageable now, you can follow the following annual plan, which should serve you here as a small cheat sheet for a rough overview of your activities over the year. In addition, ask experienced beekeepers and / or in associations for advice, develop a sense and observe especially your bee colony with regard to the tasks ahead of you during the months to bring your bees through the year species-appropriate. Since winter feeding lays the foundation for the following bee year, our plan begins in August. Do not forget that the weather of the respective seasons - unlike the length of daylight - can vary from year to year depending on the region. So it may happen that your activities also deviate slightly in time from the plan presented here. You are welcome to read about the bee colony in the course of the year in parallel from page 70 onwards and keep to the factors described there.

August:

- Weekly inspection of the hive
- Artificial swarming
- Varroa treatment (with formic acid)/TBE
- Winter feeding: a mixture of sugar solution and regis-
tered nectar is best

September:

- Weekly inspection of the bees
- Completion of winter feeding: about 20 kg of reserves
should be stored
- Re-check for increased mite infestation, if necessary
touch up with oxalic acid preparations
- Re-pollinate old colonies
- Estimate honey stocks
- Attach mouse protection at the flight hole
- Reduce flight holes and remove food remains on the
outside of the hive: protection against predation.
- Sort out honeycomb

October:

- Inspection of the bees every two weeks (no more major
brood chamber inspections)
- Check if feed quantity is sufficient
- Narrow flight hole

- Checking the correctness of the colony
- If necessary, replace queen
- Wax processing; melting down of old combs
- Cleaning and disinfection of old frames
- Stirring, inoculation and filling of honey (prepare for Christmas business)

November:

- Routine sightings from the outside
- Adapt brood chamber to colony strength in view of the coming spring
- Ensure quiet wintering: check the stability of the cane, prevent disturbance of the surface by surrounding branches, twigs, etc.
- Checking for damage after storms and wind gusts
- Evaluating the year's records: Reflecting on the past months
- Clarify melted wax
- Candle production
- Preparation for Christmas/Advent business

December:

- Last opening of the hive: residual mitigation of the comb alleys with oxalic acid dihydrate solution, document treatment in the stock book with date, colony number and quantity (use solution only once in winter)
- External controls
- Candle casting with real beeswax
- Preparation for the Christmas markets

January:

- Rest in the hive
- Weekly checks for external damage to the hive
- Cleaning and repairing tools and equipment as well as frames, hives, center walls, etc. to keep them in good condition.
- Obtain information through online beekeeping courses, non-fiction books and lectures.
- Obtain necessary materials for the coming bee year

February:

- Weekly checks for external damage to the hive
- Prepare frames and hives, mend or build new ones
- Removal of winter deadfall from the soils
- Remove dead colonies immediately and melt down combs

March:

- Observe flight hole: When do bees fly out? Is the colony already caring for the brood?
- When warm weather sets in, cautiously perform initial feed checks and replenish supplies if necessary
- Systematically control the adaptation of the brood chamber: Enlarge brood nest in a controlled manner
- Unification of orphan or weak peoples
- Continue to remove dead peoples

- In the case of a two-cage overwintering, remove the - now empty - lower brood chambers.
- Cleaning and disinfection of used frames
- Finish remaining frames and hive parts
- Prepare material for the upcoming season

April:

- Check feed stocks at the beginning of the month: about five to eight kilos should be available to protect against possible cold spells
- Check weekly for queen cells
- Set up honey chambers in the middle to end of the month
- Insert drone frame in the middle to end of the month

May:

- Check bee colony weekly
- When swarming starts, remove bees to create new smaller scions (cupping).
- Alternative: Let bees swarm and capture formed swarm cells
- Cut out drone frame every two weeks
- Timely extensions of the honey chamber
- Rearing the queens
- First honey harvest, also to prevent swarming

June:

- Forming offshoots: large colony development favors multiplication
- Check for queen cells at least once a week
- Cut out drone frame every two weeks
- Rearing the queens
- Bringing the mating box to the voucher station
- Second honey harvest

July:

- Check for queen cells at least once a week
- Continue cupping
- Prepare last honey harvest
- Control mite load and treat parasites
- Check feed and brood level of the colonies, adjust bee quantity by brood removal
- Create new colonies with bred queens

And now - off to the hive!

This book could give you a detailed insight into beekeeping and the anatomy and physiology and behavior of the honey bee. Nevertheless, you are only at the beginning of your career and at the beginning of infinite knowledge that you can achieve as a beekeeper to constantly improve and do justice to your bees. Take advantage of all the offers that your beekeeper association provides: Continuing education, training or access to the association's own library.

Getting a beekeeper mentor is an absolute recommendation. An experienced beekeeper at your side can save you many beginner's mistakes and thus ensure the fun and joy of your own bees.

If your hobby grows and flourishes to the point where you are thinking about selling honey for commercial purposes, please also remember that this falls under the Food Act and you have to meet certain requirements. Educate yourself further and acquire a certificate of competence or expertise. Your beekeepers' association can provide you with competent assistance here.

And last but not least, never forget that beekeeping should be fun. Enjoy your bees and approach your work with curiosity and interest. In the end, nothing is more beautiful than a healthy, strong bee colony - and the honey will taste even better to you than it already does, because you made it completely yourself!

(With the help of your bees, of course!)

List of technical terms

A

- The young colony that the beekeeper has taken from another colony in order to avoid swarming. A scion usually consists of several frames with center walls in an empty hive, to which several brood combs are added. Attached bees are taken over as well.

- Alarm pheromone: This pheromone is emitted by the guard bees when the colony is attacked. Thus, the other bees are alerted and the enemy is marked by the pheromone.

- Old colony: A colony that has successfully overwintered for at least one year. Also called economic colony.

- American foulbrood: bacterial bee disease caused by *Paenibacilus larvae larvae*. A notifiable animal disease.

- Nurse bees: Worker bees between 4 and 10 days old that take care of brood care.

- Apis mellifera: the scientific name of the honeybee.

- Worker: numerically the most powerful type in the bee colony. Female, performs all work in the hive except egg laying (queen) and mating (drones).

- The first crop of the year, introduced in spring, is used for the development of the colony.

B

- Construction bee: worker bee responsible for the construction of honeycombs. Characterized by the active wax gland on the abdomen.

- Penciling: The queen lays eggs, called pens, in the prepared brood combs. The process is called brooding.

- Hive: The dwelling of the bees, provided by the beekeeper. The magazine hive is the most common form.

- Bee: The bee colony in its entirety, together with the stores and the comb, is called the bee. This makes it clear that the entire colony works as a unit.

- Bee bread: mixture of pollen, unripe honey and digestive secretions of the bees, stored at the edge of the feeding wreath.

- Queen bee: female bee that is the only bee in the colony that is fertile and can lay eggs. Provides for all the offspring of the colony.

- Bee pasture: sum of all sources of pollen available to bees.

- Brood: the offspring of the bee colony. This includes eggs, larvae and pupae.

- Brood hive: Form of the offspring, formed by the beekeeper. Contains brood in all stages: eggs, larvae, pupae.

- Brood nest: Core of the bee colony. This is where the brood is raised. The place where the queen is usually found.

- Brood chamber: The part of the hive where the brood nest is located.

- Brood comb: The part of the comb, in larger colonies there are several combs, which the queen uses to lay eggs.

C

- Chitin: Polysaccharide of fungi, arthropods and insects. Forms structure and stability in the exoskeleton.

D

- Drone: Male bee that develops from an unfertilized egg. Its only job is to mate with a queen of another colony. The drone dies in the process.

- Drone frame: A frame prepared in such a way that the bees raise mainly drone larvae in it.

E

- Induction: A queen placed in the colony by the beekeeper. The process is called induction.

- Exoskeleton: the external skeleton of arthropods. Opposite to the internal skeleton of vertebrates. Formed from chitin.

F

- Fat body: organ comparable to the liver. Among other things, necessary to synthesize wax.

- Flight hole: the entrance and exit opening of the hive. With a few exceptions, it is always open so that the bees can fly in and out.

- Feeding ring: direct feed supply in the immediate vicinity of the brood comb. Serves the direct supply of the queen and the brood.

G

- Royal jelly: also called head gland secretion, royal jelly. Produced by nurse bees to feed the queen and the larva that matures into the queen. All bee larvae get this feeding juice in the first 3 to 4 days. After that only the queen.

- Gemüll: A type of cane inspection in which the remains on the ground plate are inspected for health diagnosis and general cane inspection.

- Venom sac: Located in the posterior part of the abdomen, contains the venom and is part of the sting apparatus. Empties when stung. Once emptied, venom is no longer replenished. The sting apparatus, including the venom bladder, is torn out when a mammal is stung, through its thicker skin. The bee dies in the process.

H

- Hemolymph: Blood equivalent of insects. Transports nutrients, hormones, heat and breakdown products.

- Honey bladder: a kind of crop at the end of the esophagus. Serves for honey storage and transport. Honey from the honey bladder can be released to the outside through the proboscis.

- Honeydew: excretions of leaf and bark aphids. Mixture of sugar and water.

I

- Imago: adult insect.

- Inoculation: Used to initiate the crystallization process in certain types of honey.

J

- Young people: people that have emerged in the course of the current calendar year.

K

- Kairomones: messenger substances that are not pheromones. Used for communication between different species. Bees' kairomones are picked up by mites, for example.

- Queen substance: secretion of the mandibular glands of

the queen. Acts as a pheromone and promotes colony cohesion.

- Artificial swarm: Possibility of forming an offshoot of one's own colony. Swarming is simulated in order to guarantee a new start of the colony that is as close to natural as possible. No brood and no other combs are taken over from the old colony.

L

- Empty frame: Frame without center wall and without wire. Used by the bees as a natural construction option.

M

- Magazine hive, short magazine: Consists of a floor, at least one frame and a roof. Suitable for housing bees. After removing the roof, each honeycomb can be removed individually from the top.

- Center wall: An artificially made beeswax plate that is hung in the frames to help the bees build the combs.

N

- Re-creation cell: after the loss of a queen, the colony can, if there are larvae in the colony, which are not older than three days, raise a new queen. The larvae combs are converted into replenishment cells.

- Natural construction: is carried out by the bees in empty frames. It is carried out without a predefined central wall.

It is considered the most original form of honeycomb construction.

- Nectar: sugary liquid of the flowers of most plants. Serves to attract bees, which use the nectar as food.

O

- Orientation flight: The first flights that the bees make out of the hive. Either after transfer or after hatching, as soon as the hive bee becomes a flight bee.

P

- Pheromones: messenger and attractant substances of the bees that are released to the outside and serve for hive communication.

- Pollen: also called pollen. Serves for sexual reproduction of plants, spread by bees by flying around. Collected by the bees as a source of protein.

- Pollen bread: see bee bread

- Propolis: putty resin of the bees. Collected from the bud scales of some tree species, especially poplar, birch, alder and chestnut. Saliva and wax are added. Serves to seal the stick against drafts and moisture. Has an antibacterial effect.

- Cleaning bee: bee directly after hatching. The tasks are limited to cleaning the hive, especially the brood combs.

Q

- Quacking: Hatching queens in the queen cells make this sound to check whether there is still an old queen in the hive.

R

- Cleaning flight: Bees do not defecate in the hive. To get rid of excrement, they undertake cleaning flights.

S

- Gathering bee: Bee responsible for the gathering of tracht. The last stage of bee development.

- Tail dance: complex form of communication. Bees inform their hive mates about the quality and quantity of the source of the pollen.

- Swarming: Swarming, the formation of a swarm of bees, happens when the colony becomes too large for the current dwelling. Part of the colony moves out and looks for a new home.

- Swarm cell: also queen cell. A brood cell in which a new queen is raised.

- Smoker: utensil that produces smoke, which is directed into the hive to calm the bees.

- Summer bee: bee born in summer, incubates between March and August. Survival time is about 6 weeks.

- Varietal honey: varietal honey from the trait of individual plant species.

- Track bees: scouts of the bees. Search for new sources of honey or dwellings during swarming.

- Hive bee: all workers in the first 20 days of their life. During this time, the hive is not left.

- Hive map: The bookkeeping of beekeeping, in which everything is noted that concerns the respective bee colony.

- Chisel: a typical tool used in beekeeping. Small metal chisel for loosening stuck frames and frames and for cutting out natural construction.

T

- Tracht: Entirety of all supply from pollen, nectar and honeydew. The nutritional basis of the bee colony and the basis of a good honey harvest.

- Trachtlücke: The absence of enough grapevine to sustain the colony. Often in rural areas, when the mass harvest of agriculturally grown crops is over.

- Trophallaxis: social feeding. This is how bees exchange the contents of their honey bladders. This is not only an exchange of food, but mainly of pheromones and information. The queen substance is distributed in the colony by trophallaxis.

U

- Repositioning: Replacement of the queen from the colony. Silent re-pollination occurs with an old queen, even without the beekeeper's intervention.

V

- Varroa mite: Varroa destructor: parasite of bees, nowadays present in almost every beekeeping.

W

- Honeycomb: Bee nests consist of several honeycombs. Honeycombs have several tasks. They are used for honey storage and for safe brood rearing. Honey combs are used for honey storage, brood combs for brood rearing.

- Weisel: another word for the queen bee.

- Winter bee: Bees that are incubated between August and September. They live until the next year March/April and thus have a significantly longer life expectancy than summer bees.

- Economic colony: see old colony

Z

- Frame: part of the magazine hive. In the frames hang the frames with the combs.

- Adding cage: A small cage for the queen bee to facilitate

adding to a new colony and avoid killing the queen by the
colony.